Research and Revolution

Research and Revolution

Science Policy and Societal Change in China

Richard P. Suttmeier
Hamilton College

Lexington Books
D.C. Heath and Company
Lexington, Massachusetts
Toronto London

Library of Congress Cataloging in Publication Data

Suttmeier, Richard P.
Research and revolution.

Bibliography: p.
1. Science and state–China. 2. Science–Social aspects–China. I. Title.
Q127.C5S86 338.4'7'50951 74-8108
ISBN 0-669-94565-x

Published simultaneously in Canada.

Printed in the United States of America.

International Standard Book Number: 0-669-94565-x

Library of Congress Catalog Card Number: 74-8108

For Christopher and
Harriet Suttmeier
and Merle

Contents

List of Figures

List of Tables

Preface

The inspiration for this study goes back more than ten years, when as a fresh recipient of my bachelor's degree, I went off to Hong Kong as a Dartmouth Project Asia Fellow to teach the history of science at Chung Chi College of the Chinese University of Hong Kong. The two years spent at Chung Chi provided a unique opportunity to be a participant in and observer of the process of Chinese scientific development. As participant, I was engaged in my small way in the cross-cultural exchange of ideas that is so much a part of that development. As observer, I learned a great deal from my senior Chinese colleagues. The professional lives of most of these had necessarily been touched by the great cultural struggles of twentieth-century China involving the reconciliation of Chinese learning and Western learning, and particularly, modern science and traditional culture.

Perhaps the most formative impression, however, was made by my students, particularly the science majors. I was consistently struck by the influence of utilitarian values in motivating them to begin to study science, and by the fact that for many, the utilitarianism had given way to values of "science for its own sake" by the time they graduated. In retrospect, it is clear that what I was witnessing was the powerful socialization into a unique social system that is a part of scientific training—a genuinely universal trait of modern science. The students and colleagues of Chung Chi, who made the stay there so worthwhile, are too numerous to mention individually. I trust they know my appreciation.

My interest in Chinese scientific development grew somewhat more systematically after I entered graduate school at Indiana University. At Indiana, I was able to work out an integrated program which combined work in political and administrative development, science and public policy, Chinese politics, and history and philosophy of science. I am grateful for the support provided by Indiana University in various forms and for the freedom and opportunity it provided to pursue these somewhat diverse fields. In particular, I am grateful for the guidance provided during that period by William J. Siffin, Lynton K. Caldwell, and Ilpyong Kim.

The opportunity to compose my thinking and ongoing research on the interrelated subjects of science policy, societal change, and scientific development in China into the present work was afforded by the Technology and Development Institute of the East-West Center during 1972-73. I am grateful for the generous support and gracious hospitality provided me in Honolulu by TDI. In particular, I would like to thank Dr. Hahn-been Lee, the former director of TDI, for his kind invitation to spend a year at the Institute as a visiting fellow.

Various people have read the manuscript or parts thereof. These include Alok Chakrabarti, Richard Gable, Everett Hagen, Young-whan Hahn, Eugene Lewis, Leo Orleans, John Richards, Maurice Richter, and William Snead. For the valuable comments and suggestions they provided, I am most grateful. For the remaining errors in fact or interpretation, I am of course responsible.

My appreciation also extends to those who typed through the various versions of the manuscript. The bulk of the work was done by the TDI secretarial staff. During the final preparation of the manuscript for publication, the efforts of Mrs. Esther Jusza and Mrs. Bonnie Bowie were most helpful. The assistance provided by Hamilton College in supporting my research during the summer of 1971 and in the final stages of preparing the manuscript is also much appreciated.

Finally, for temperance during the manic phases and encouragement during the depressive, the support from my wife Merle has been indispensable.

Introduction

. . . with the manufacture of power a new epoch began . . . [and] in many ways the new epoch must open as an era of destruction. It must from its very nature destroy many of the conditions which give most interest to the history of the past, and many of the traditions which people hold most dear. . . . There must be a great destruction, both in the physical and in the intellectual world, of old buildings and the old boundaries and old monuments and, furthermore, of customs and ideas, systems of thought and methods of education.

—Elting Morison

This book is about science and society in the world's most populous country. It ambitiously attempts to reach three audiences. First, it is intended to reach the community of China scholars who are only recently beginning to pay attention to the role of science and technology in the development of the People's Republic. Second, it is addressed to development scholars who have become not only increasingly concerned with the Chinese model for development but also increasingly interested in the general relationships of science, technology, and development. Finally, the book is for the more general audience of those who are troubled by the problems of science and society and would welcome the comparative perspective on these problems that a description of Chinese experience might offer.

The book is written at the end of an era of Sino-American relations. It is of the "outsider-looking-in" genre, typical of most American scholarship during that era. It is therefore tentative. It represents an attempt at an interpretive summary of twenty years of Chinese science and technology policies and programs. It attempts to look beyond the individual trees to see the patterns of a unique forest.

"Science and society" is a phrase that is both evocative and terribly general. It evokes a plethora of interpretations and connotations. The use of the conjunction between the two nouns in the phrase implies some relationship between "science" and "society." In a very general but profound sense, science and society can be thought of as a nearly all-inclusive category under which the great body of human experience—in terms of *knowledge and action*—can be aggregated. This knowledge and action connotation is so broad that its usefulness for our purposes is limited. It does however have a significance that is worth noting.

That significance is manifest when we broaden our context to include questions of change, development, or modernization. The inclusion of questions of change immediately raises the empirical question of the immutability of the knowledge-action, science-society relationship. Empirically, it does appear that over the course of world history the relationship has not been a constant one and that over time action has increasingly been a response to knowledge. This of course is an article of faith held by students of societal change who go so far as to argue that "modern" society is one in which the way and spirit of science predominates.

As if to illustrate their own point, those concerned with modernization have used this straightforward piece of empirical *knowledge* to recommend *action* programs designed to foster science and the expansion of knowledge in countries seeking to modernize. To the extent that these actions are undertaken by governments, we can speak of them as science policies. They can be thought of as *developmental* science policies in that they seek to develop modern science in the interest of modernization in environments that were not previously characterized by high levels of scientific activity.

Developmental science policy, however, jumps ahead of the more fundamental question of causal relationships between science and society within the context of change. If one experiences change in science, can one then anticipate change in society? Conversely, what kinds of change in society would lead one to expect change in science? More deterministically, does science *cause* societal change or does societal change *cause* scientific change? A minimal prerequisite for dealing with these questions is to clarify some concepts. This will be attempted in the first chapter. Here it is sufficient to note three points.

First, developmental science policy appears to be built on the premise that science and technology are positive forces for societal change. Second, in spite of this, there does not seem to be unanimity on interpreting the science-society relationship in this way. Marxists particularly, but some non-Marxists as well, see the science-society relationship as far more complex and, if anything, point to the dynamics of societal change as a force in scientific change.

Finally, without having to settle on a firm science–society interpretation, one can see an approach to science policy other than the developmental approach. This alternative we will call *integrative* science policy. Integrative science policy assumes that the science-society relationship is exceedingly complex and that the lines of causation run two ways or in cycles. The goal of integrative science policy is not primarily the development of science (although that may be one goal). It is, rather, to modulate, coordinate and balance scientific change and societal change. Integrative science policy is a relatively new phenomenon that is now being given serious attention in the techno-scientific societies of the industrialized nations.[1] Although our interest here is primarily with developmental science policy, an underlying concern of this inquiry is the relevance of integrative science policy for those nations seeking to modernize.

We have set out some of the general issues of science and society to signal the reader that Chinese efforts to foster a national tradition of research and innovation are part of the larger global drama of defining the science-society relationship. The audience for this drama as well as the number of its participants will grow as we approach the twenty-first century, and China's role will certainly increase. The magnitude of this role warrants our interest, and we will want to return to it in the concluding chapter, after we have reviewed the last twenty years of China's experience. First, however, an exploration into definitions and perspectives is in order.

Note

1. Organization for Economic Cooperation and Development (OECD), *Science, Growth, and Society: A New Perspective* (Paris: OECD, 1971).

1 Research, Innovation, and Developmental Science Policy: A Perspective

Once there is a technical need in society, such need drives science forward better than 10 universities.

—Frederich Engels

Introduction

A recent report from the Organization for Economic Cooperation and Development (OECD) contains the simple but pregnant phrase "technological innovation is more than R&D."[1] In a world that places great value on science and technology increasingly for instrumental purposes, it is worthwhile attempting to bring forth the meaning and implications of this phrase.

From Detroit to Delhi, Peking to Paris, and Cambridge to Cairo such words as "research and development" "science and technology," and "invention and innovation" are exchanged not only among scholars but among politicians, bureaucrats, industrialists, and agriculturalists at a rate that would have been unimaginable 100 years ago. The reason is clear. New knowledge, artifacts, and processes, which also would have been unimaginable 100 years ago, have revolutionized man's power over nature and over other men. In the latter part of the twentieth century, those individuals, organizations, and nations wishing to increase their economic productivity, enhance their wealth and/or quality of life, or extend their power will increasingly be preoccupied with these nostrums.

The attention now given to science, research, and innovation has led to a great amount of attention to *science policy*. The policy unit generally has been the nation-state, and science policy has been understood therefore as "a deliberate and coherent basis for national decisions influencing the investments, institutional structure, creativity, and utilization of scientific research."[2] As such, science policy has actually meant policy for technology and technological innovation as well as for science. Science policy has usually been thought to include policy for science—the "need to manage the scientific and technological resources of a country as a whole"—and science in policy—"to understand more clearly the influence of new scientific and technological developments on other elements of government policies."[3]

There is a certain ambiguity about the notion of science policy that stems from the question of whether discovery in research and technological innovation are essentially unpredictable phenomena. If so, science policy cannot serve as a *plan* for discovery or innovation. On the other hand, productive research and innovation are not unrelated to the availability of certain kinds of resources and to factors of social organization. While science policy may not be an actual plan for discovery, it can nevertheless serve as a program for creating an environment conducive to research and innovation. In this sense, policy is an instrument to reduce randomness, and as such reflects choices and subsequent enforcement of decisions about the allocation of resources and the modes of organizing research and innovation systems. Such choices are made within the context of overall national objectives.

There is a growing awareness of the special importance of science policy in low-income countries.[4] The scarcity of human and material resources for science and technology imposes constraints on the development of science and technology that call for a rational policy of resource utilization. But when it comes to science and technology, what constitutes a *rational* policy? Before answering this question, it may be helpful to examine certain characteristics of science and technology and research and innovation as they exist in industrialized countries and to contrast these with the status of science and technology in low-income countries.

The phenomena that we call science, technology, research, and innovation are characterized both by individual intellectual activity and by the social interaction of individuals. Sociologists of science, since the pioneering work of Robert Merton, have been particularly sensitive about mixing conceptual and social aspects in the definition of science.

In defining *science*, Norman Storer, for instance, sees the ideational side of science as a body of systematic knowledge arrived at by using accepted physical and mental techniques of investigation. But Storer also recognizes that this body of knowledge and the techniques of investigation are not independent of an "organized activity of men and women who are concerned with extending man's body of empirical knowledge through the use of these techniques."[5] As an organized social activity, science exhibits characteristics of a social system or a community. Such characteristics include social relationships that are conducted according to shared values and norms, mechanisms of social control, and the exchange of a commodity that is uniquely valued by the community.[6] It is clear that, in the sense described above, science can also be thought of as an institution, or

> . . . a normative pattern, conformity with which is generally to be expected, and failure to conform with which is generally met with the moral indignation of those individuals who belong to the same general social system and are aware of the failure . . . or by some other sanctions from that source.[7]

Similarly, science, like institutions generally, has a set of identifiable social roles and a system of recruitment and socialization to introduce new members into these roles. Finally, an institution possesses what alternatively has been called a "doctrine" or "code." Doctrine is understood as the expression of the values the institution represents—in terms of its goals and the norms governing the behavior appropriate for attaining those goals. Thus, doctrine concerns a statement of means and ends and in this sense is similar to what Martin Landau has called a code.

> For means-ends relationships, it must be always be remembered, are problem-solving devices, and all the societies have their trusted recipes for meeting the untoward. Indeed, when it is said that members of a society are "socialized," what they learn are problem-solving programs—sets of formulae, maps, maze-ways, or "codes" by means of which they make sense of the life-space in which they find themselves. They know what stimuli to respond to, when and how to respond, and under what conditions they may alter or modify responses.[8]

Science as a social system or institution can be understood, then, as an organized human activity characterized as having its own "problem-solving devices" and "trusted recipes," which are integral to the normative characteristics of science.

Technology has assumed a variety of meanings from hardware and artifacts to "the totality of methods rationally arrived at and having absolute efficiency in every field of human activity."[9] The Harvard University Program on Technology and Society defined technology as "the organization of knowledge for practical purposes."[10] As with science, technology has both a social and a non-social dimension. The latter would certainly include hardware—a wide variety of machinery and artifacts—as well as an ideational component. It is the social dimension of technology, however, that will most concern us. Can we speak of technology as a social system or institution in the same way as we spoke about science? The answer is a qualified yes.

Technologists like scientists have recruitment and socialization mechanisms for introducing new people into technological roles. Likewise technologists have internalized problem-solving devices and trusted recipes. Technologists like scientists have such formal manifestations of institutionalization as professional societies and journals, although these appear less influential for technologists than for scientists. Here may lie the main social differences between science and technology.

Thomas Kuhn has noted the influence of paradigms in science for setting the means-ends calculus of the community. Kuhn notes that such achievements as Newton's *Principia* and *Optiks*, Franklin's *Electricity*, Lavoisier's *Chemistry* and Lyell's *Geology* were "sufficiently unprecedented to attract an enduring group of adherents away from competing modes of scientific activity," and were "sufficiently open-ended to leave all sorts of problems for the redefined group of practitioners to resolve."[11] Goal-setting, then, is done under paradigmatic guidance and has both an intellectual and sociological dimension.

Under periods of normal science—that is, when the existing paradigm holds unquestioned allegiance—the intellectual task involves the progressive application of the theoretical content of the paradigm to cover more and more phenomena. Paradigmatic guidance also involves social guidance from the scientific community through the socialization process occurring during advanced scientific training, through the community's control of scientific journals, and more recently through the community's influence in the allocation of research funds. In a similar fashion, the choice of means that contribute to science as method reflects the influence of the paradigm. Witness for example the intellectual training and socialization that occur during the long hours spent by a graduate student in the sciences learning experimental techniques. The main intrusion on the autonomy of science in the choice of means has been that provided by the producers of new scientific instruments, machines, and gadgets. In the main, however, the setting of means and ends in basic science has been a function of the intellectual and social forces within the scientific community itself.

Although it has been suggested that in some ways the technological state of the art functions as a paradigm for technologists,[12] it would appear that this is true more for the setting of means than of ends. Indeed, it appears that the setting of ends for technology is not generally a function of factors internal to the social system of technology but rather of factors in the environment, whether they be those of the market, those of national defense, or those diffuse factors relating to quality of life.

Research and Development (R&D) is a widely used term. Definitions of the elements of this term have become standardized.

> *Basic research* is that type of research which is directed towards increase of knowledge in science. It is research where the primary aim of the investigator is a fuller knowledge or understanding of the subject under study, rather than, as in the case with applied research, a practical application thereof.
>
> *Applied research* projects are those . . . which represent investigation directed to discovery of new scientific knowledge and which have specific commercial objectives with respect to either products or processes.
>
> *Development* refers to technical activity concerned with non-routine problems which are encountered in translating research findings or other general scientific knowledge into products or processes. It does not include routine technical services to customers or other items excluded from definition of research and development above.[13]

These definitions have been used in estimates of the allocation of expenditures or manpower among the three activities. Analysts, more concerned with *processes* relating to the origins and diffusion of knowledge, have often written in terms of invention and innovation. *Invention* refers to an "idea of how science and technology could be applied in a new way."[14] *Innovation* refers to actions having the effect of "bringing invention to its first commercial use."[15] Finally, diffusion of innovation is taken to mean "the spread of the use of the innovation amongst its potential population of users."[16] With these definitions we have again introduced terms that refer to ideas on one hand and to social phenomena on the other.

It will be helpful to examine more carefully the relationship between social and ideational concepts, precisely because the Chinese have been so conscious of this relationship themselves. What follows then is an attempt to synthesize some scattered views on science, technology, developmental science policy, and the elements of the innovative process into a framework for comparative analysis. The intent is to place our subject—Chinese efforts to foster science and innovation—in a perspective that incorporates both the social and ideational components in order to facilitate the application of insights developed elsewhere to the Chinese case.

Organization and Innovation

In terms of social impact, the outstanding feature of research in the modern world has been the marriage of scientific and technological creativity with the resources and purposes of large organizations, including governments. Organizations have taken Bacon at his word—knowledge *is* power—and have searched for formulae for fostering the growth of science and technology and their uses. On the other hand, as scientific research has expanded in scope, it has increased in cost, and science has increasingly come to rely on the resources that only large organizations can supply.

The resulting, uneasy situation has raised fears and false expectations that generally have been a function of a lack of empirical information. At one extreme, one can still discern the position of the scientific purist. This position, which has lost many of its adherents since some of its more dire predictions have not obtained, is that scientific advance is a product of the creative interaction of the members of the scientific community and that science itself has little to do with practical affairs. The acceptance of money from large organizations for research is both unnecessary and dangerous—unnecessary because science has nothing to do with funding and dangerous because it exposes the means-ends calculus of science to intrusion from outside. Once scientific ends are set by outsiders one is no longer doing science. This position might be called the "pure science ideology."

At the other extreme are the views of hard-headed men of affairs, the representatives of large organizations. These men are impressed by the wonders wrought by R&D and feel they too must have science incorporated within their organizations or governments. It may be, however, that these men are in fact interested in technological innovations and quick payoffs for funds invested in research. They show little interest in science itself and grow increasingly impatient with "eggheads" who are squandering the organization's funds in the pursuit of narrow interests. This might be called the "quick-fix ideology."

Between the pure science ideology and the quick-fix ideology are gradations of other research ideologies. It is important to recognize the existence of ideologies of science, because they represent the values, attitudes, and expectations of the actors in the worldwide R&D drama, and their ideologies have action consequences. One of the tasks of any research effort on research is to provide empirical evidence that will help insure a better fit between ideology and reality.

Science, it has been said, can be thought of as an institution or social system, although traditionally a relatively autonomous system of social organization. The individual scientist is the keystone of the institution. He both makes the institution viable and in turn is most evident in the unique exchange mechanism that characterizes science, a mechanism that might be called the information-recognition (i-r) exchange. The scientist as an individual is engaged in various creative intellectual processes that in the best of circumstances lead to discovery. What makes science an institution, however, is that the scientist is both eager to, and expected to, submit his discovery or findings to the community of his peers. The community reviews his work, and hopefully will extend to him the most prized commodity within science, colleague recognition in the form of citation of his work by others. This phenomenon, which has been the subject of sociological analysis, is explained by the norms of the scientific community that have been internalized by its members.[17] Scientists are enjoined by these norms to participate in the information-recognition exchange both as suppliers of information and as critics and recognizers. Sociologists are not alone in recognizing this behavior. A scientist, recently reflecting upon his profession, noted that:

> . . . scientific loyalty is not just towards a prestigious professional group but to an ideology . . . Being a successful scientist is not just winning prizes, it is having other scientists cite your work. To give this up is worse than losing caste; it is to give up one's faith and be made to worship foreign idols.[18]

The implications of this ideology are significant. The first is the importance of communication in science, particularly through professional journals. Second, it implies that the basic motivations of scientists comes not from generalized societal recognition and reward but from recognition from a professionally competent peer group. A final implication is that inasmuch as competence or expertise resides within the professional peer group, there is little justification for intrusions from outside the group that in any way threaten the autonomy of the group in matters pertaining to that competence. The possession of expertise justifies autonomy, and expertise can best be exercised under conditions of autonomy.[19] Traditionally at least, the complexities of social organization in science were limited to the relatively autonomous colleague community. This organization was not complex and was highly compatible with, and supportive of, individual creativity.

It will be recalled that technology is significantly different from science in that rather than being a "closed" system in terms of means and ends, it is concerned with creatively devising means for ends set by others. One scientist has noted that "the technologist's prime responsibility is towards his employer, his customer or his patient, not to his professional peers."[20] Technologists have been described as far more deployable than scientists in that their expertise can be directed toward the solution of problems that have been identified by others.[21]

A recent study of communications patterns in technology by Donald G. Marquis and Thomas J. Allen sheds additional light on the significant sociological differences between science and technology. Marquis and Allen discovered that although technology has developed a journal literature, it is different from science in a fundamental way. Its literature is not cumulative. Citations are fewer and are often to the author's own work. In general, publication is less highly valued than it is in science—"the technologists' principal legacy to posterity is encoded in physical not verbal structure."[22] Communication within technology is primarily oral—between co-workers within their organization, and between technologists and customers or vendors outside the organization. The Marquis and Allen findings are consistent with Price's remark that whereas science is papyrocentric, technology appears to be papyrophobic.[23] One of the implications of this characteristic of technology is that inasmuch as interpersonal communication is crucial to technology, factors of social organization that either facilitate or inhibit communication bear directly on technological progress.

Factors of social organization take on even greater importance when we shift the level of discourse from certain sociological characteristics of technologists to the process of technological innovation. Innovation, it will be recalled, involves those actions necessary to bring an invention into use. In this sense, the relationship between technology and social organization becomes crucial, and indeed in some cases, a change in social organization itself might constitute a technological innovation. Louis Morton notes a case from military history in which changed organization made possible a new utilization of an old artifact, and thus resulted in a genuine technological innovation.

> The pike is nothing but a thrusting spear, not much different from the sharp stick of primitive man. Yet in the hands of the Greeks it was powerful enough to turn back the great armies of the Persian empire. The reason lay in the formation they created to use the pike—the phalanx. The Romans used the same basic weapons as the Greeks, . . . but employed them differently. The result was the Roman legion, the most successful and remarkable formation devised by any society until modern times.[24]

Other cases from military history illustrating the relationship between artifact and organization are recounted in the fascinating book by Elting Morison.[25] Morison's discussion of the development of continuous-aim firing aboard U.S.

naval vessels indicates how the adoption of a new artifact in the end changed social organization on board ship. But the continuous-aim firing case, as well as the case of the introduction of the steam-powered vessel Wamponoag, also demonstrate another crucial aspect of the technology-organization relationship—that is, the cultures and operating principles of organizations are often premised upon existing technologies. Technological innovation can have the effect of upsetting organizational codes and is therefore resisted in spite of the advantages offered by the innovation for meeting organizational goals in the long run. Indeed, Morison's accounts raise the question of whether one can seriously talk about technological innovation at all without first talking about the capability of organizations to be innovative and about the forms of organization and modes of organizational behavior that are most innovative.

In a recent essay, Victor Thompson has raised a number of interesting questions about the relationship between innovation and *bureaucratic* (or *monocratic*) organization. He suggests the two are incompatible. The modern bureaucrat, according to Thompson, is faced with a system of organizational accountability and a reward structure that promotes a conservative avoidance of risks, and risk, it is thought, is an essential ingredient in innovation. In addition, according to Thompson, bureaucracies suppress conflict, and

> The inability to legitimize conflict depresses creativity. Conflict generates problems and uncertainties and different ideas. . . . Other things being equal, the less bureaucratized the organization, the more conflict and uncertainty, and the more innovation.[26]

Thompson questions whether the essential features of bureaucracy, particularly its attempts at rationality in decision making, which is "intellectually structured in monocratic, comprehensive, problem-solving terms," are suitable for innovation.[27] Such features lead to policies, procedures, and organizational relationships that are incompatible with a high level of innovation. Rather, Thompson sees the innovative organization as "pluralistic" ("one aspect of pluralism is to have more than one criterion by which such terms as 'best' are described"), "pragmatic," "sequential, trial and error decision makers," which possess free, uncommitted resources which can be used to give full play to innovative activities ("slack").

The worst aspects of bureaucratic organization may indeed be incompatible with innovation. However, it is often difficult to point to or devise an alternative form of organization. This may be due to the complexity of innovation itself. It has been noted that

> . . . innovation is not just one simple act. It is not just a new understanding or the discovery of a new phenomenon, not just a flash of creative invention, not just the development of a new product or a manufacturing process; nor is it simply the creation of new capital and consumer markets. Rather innovation involves related creative activity in all these areas. It is a connected process in

which many and sufficient creative acts, from research through service, couple together in an integrated way for a common goal.[28]

If we assume innovation does involve all of the above, there is some question whether any one organization can be expected to perform all the above activities. The author of the quote above, Jack Morton, was associated with the Bell System, and not surprisingly he talks in terms of a "systems approach to innovation." In such a system, different organizational units perform different functions. "The essential virtue in the systems approach to innovation lies in the parts of the process and their linkages with one another, not in the sequence in which such linkages are performed."[29] The implication of Morton's experience is that no one organizational form can be specified. However, it may be possible to specify organizational traits that are consistent with the process of innovation.

Technological innovation involves uncertainty and change. It has been noted that the transfer of technology, which is part of technological innovation, is highly dependent on personal contact—it is "person embodied." The innovative organization therefore must be able to respond to these characteristics of the process. The uncertainty of innovation calls for an organization that encourages and rewards risk taking and the willingness to make a reasoned departure into the unknown. The inevitability of change associated with innovation requires an organization that can reduce the social costs of change, soften its impact, and respond creatively to it. Organizations that fear change would seem to be inconsistent with technological innovation. Finally, an innovative organization must facilitate the movement of ideas, specifically the movement of people with ideas, and the opening of channels of communication.[30]

There are a number of factors that infringe on successful communication within and between complex organizations. The first of these is the problem of specialized languages and the deeper specialized cultures and normative commitments that the languages reflect.[31] Efforts must be made to achieve a measure of commonality of language in the innovative organization. For instance within R&D departments or divisions, there is a need to overcome the specialized languages of traditional academic disciplines in order to successfully accomplish task-oriented research, which is often interdisciplinary. More important perhaps is the problem of overcoming the language barriers between R&D departments and production and distribution departments.

Spatial and organizational factors are a second group affecting communication. An interesting example of these factors in operation and a creative organizational response is offered by Morton in his account of the Bell System (see Figure 1-1). Bell Labs have responsibility for basic research, fundamental development, and development and design. Western Electric has responsibility for engineering for manufacturing and for the manufacturing itself. A crucial linkage in the innovative process is that between development and design, and

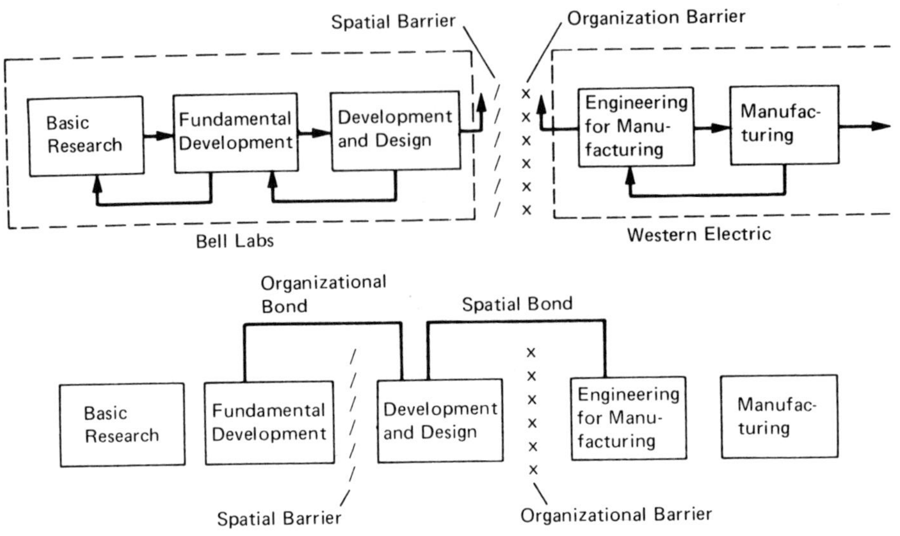

Figure 1-1. Reorganizing to Overcome Spatial and Organizational Barriers to Innovation. Source: J.A. Morton, *Organizing for Innovation* (New York: McGraw-Hill, 1971), pp. 64 and 65. Reprinted by permission of the publisher.

engineering for manufacturing. Originally this linkage was impeded by organizational structure and distance. The solution was to physically locate Bell Labs' development and design people with Western Electric. As a result, (see Figure 1-1) organizational barriers were overcome with spatial bonds, and spatial barriers were overcome with organizational bonds.[32]

A third factor affecting communication is the extent of social stratification within the research and innovation system. Status differentials exist within professions, between professions (e.g., physicists and engineers), and among professionals and workers and management. To the extent that status is threatened by communication, the protection of status among various groups may prove a significant barrier to communication.

A final factor relating to communication is motivation to communicate and engage in interpersonal interaction. This factor is related to the three above and could lead to the following kind of problem. It can be assumed that in terms of background, training, operating principles, and sense of status, the culture of an organization's R&D division will differ significantly from that of its manufacturing division. In the absence of good personal interaction it is possible that manufacturing will regard R&D as a group of impractical eggheads who are unfamiliar with, and irrelevant to, the real problems of manufacturing. Similarly R&D may feel that manufacturing is stodgy, uncreative, and unreceptive to new ideas. Manufacturing may become unwilling to acknowledge help from R&D and may, in fact, actively resist new proposals coming from R&D, as the R&D people claim. Increasingly the two divisions will become competitors for the favor of top management. Management may increasingly question the value of supporting R&D since its inventions are not being utilized (i.e., the innovation chain is broken).

This is a scenario for organizational deterioration that has been drawn by Chris Argyris and has been attributed to an absence of interpersonal competence.[33] Argyris measures interpersonal competence at a number of levels, but it will suffice here to note that interpersonal competence includes an awareness of the implications of one's behavior and a willingness to take responsibility for it; an openness in one's awareness which both "permits and encourages the reception of new information"; a willingness to experiment in manipulating an environment in order to create new information and to accept the risks involved; a willingness to help others; and an adherence to the norms of individuality (not conformity), concern for others (not antagonism) and trust (not mistrust).[34] It should be noted that interpersonal competence relates to the emotional dimensions of an organization not solely to its technical aspects. However technically competent an organization may be, without qualities of interpersonal competence, it runs the risk of sliding into inefficiencies and of resisting innovation. The corollary then is that innovative organizations must maintain a culture that sustains interpersonal competence.

The achievement of at least some of these characteristics places a heavy responsibility upon the management of innovative organizations. The concept of "entrepreneurial" management comes up more than once in the literature on innovative organizations, and, as with much that has been said above, appears consistent with Thompson's critique of bureaucracy. The OECD report cited above notes that technological innovation requires three factors: (1) a capability in science and technology; (2) market demand and the ability to perceive the demand; and (3) an agent that transforms this capability into goods and services that satisfy the demand.[35] Accordingly, the report calls for management consistent with

> "entrepreneurial" organization forms, with flexible definitions of responsibilities and large possibilities for lateral communications, capable of evaluating and responding to new—and often unforeseen—technical and market circumstances.[36]

Charles Dailey has recently attempted to characterize entrepreneurial management and to contrast it with the bureaucratic style. A comparative summary is offered in Table 1-1.[37] Worth noting about entrepreneurial styles is a willingness to seize opportunities, an orientation toward the environment of an organization, and an ability to escape a rigid code, which reflects aspirations for rationality through policy.

A particularly insightful account of management in an innovative organiza-

Table 1-1
Bureaucratic versus Entrepreneurial Styles of Management

	Bureaucratic	Entrepreneurial
Definition	Organization viewed as an impersonal machine governed by rules, not people.	Organization viewed as a living being, growing and change-seeking, infused with personality.
Origins	Excesses of industrial revolution and of spoils system in politics.	An innovative idea for a product, service, or realization of an opportunity.
Purposes of Organization	Control behavior and prevent excess.	Release human energies by presenting opportunities which call them forth.
Measurement	Measure internal occurrences, especially deviations from plan, policy, or budget.	Evaluate environmental responses to the organization's action on products.
Structural Principles	Hierarchic, derived from military and/or from mass-production models.	Process, project, and goal-oriented, derived from marketing or R & D models.
Objectives of Work	The work itself.	The effects of the work.
View of the Future	The future is a continuation of the past; desirable future is a continuation of the past.	The future is what is possible, in realizing its opportunities.
The Past	Limits the future.	Example of what can happen but not seen as a limit.
Attitude toward Change	Stability is more important than change.	The most important thing is to choose which changes will be most productive and to seek them.

Source: Charles A. Dailey, *Entrepreneurial Management*, (New York: McGraw-Hill, 1971), p. 21. Reprinted by permission of the publisher.

tion is offered by Morton in the work already cited. Morton's image of the creative manager is that of Maxwell's "Demon," an image that is consistent with Dailey's modern entrepreneurial manager. Morton's manager is seen as one who monitors the internal workings of the organization, the environment, and the interaction of the two.

> He must be alert and sensitive to changes both in his organism and its ecosystem as they interact. He will want to stimulate some of these changes himself, and to reinforce or discourage others generated by his living organism. In all cases, he wants to encourage those changes in purpose, content, and structure which will contribute most to the long-term survival and growth of his system.[38]

Thus, the manager filters and selects the "hot ideas" and organizational mutations that are most consistent with the long-term achievement of organizational goals.

Science, Technology and Innovation

It should be clear by now why "technological innovation is more than R&D." The end product of development at best is a prototype that must somehow find its way into production and routine use if innovation is to occur. This we have seen involves complex social processes. But if innovation is more than R&D, it should be emphasized that R&D is still an important part of the process in modern economies, and for this reason, industry and governments are taking the lead in sponsoring the establishment of research departments and institutes and committing relatively scarce resources to these institutions. The introduction of research activities into the process of large organizations has not been without problems, however. Kornhauser, for instance, has studied the problems of scientists in American industrial firms and has discovered that professional scientists are in conflict with management over such issues as goals for research, control over research work, incentives, and influence over how the results of research should be used.[39] These conflicts can be expected precisely because of the institutional nature of science, the influence of paradigms, and the importance of the information-recognition exchanges. Significantly though, Kornhauser and others[40] have found that accommodations and adaptations that occur between scientists and large organizations mitigate the conflict and that these can be mutually beneficial. Leo Meltzer has found that scientific productivity is greater when some autonomy is sacrificed to organizational constraints in the interest of greater material support than in settings where there is a very high degree of autonomy, but low degree of resources, or in settings where resources are plentiful but autonomy is low.[41] The worst fears of the scientific community about the possibilities of accepting organizational resources, and with them organizational constraints, have proven over the years to be groundless as a matter of *necessity*. Rather, the opportunities for doing high-quality research in organizational settings have been *contingent* on a willingness to seek mutual understanding and adjustments and on enlightened management.

There is also evidence that research, particularly applied research, in organizational settings might actually benefit from factors of the organizational setting other than resources. The immediacy of problems, the necessity for interdisciplinary effort and the pressure to attack problems in new ways can be seen as contributing to innovative and creative research. Gerald Gordon and Sue Marquis report that medical sociologists are more innovative in marginally academic settings than in purely academic settings. The explanation given seems applicable to the physical sciences as well. Scientists within the autonomous scientific community, free from the constraints of organizational goals, are in periods of normal science under paradigmatic guidance. The reward pattern of normal science induces conformity rather than originality in terms of the selection of means and ends. However, in the marginally academic setting—that is, where organizational goals are present and the consequences of research can be assessed in terms of those goals—the situation changes. According to Gordon and Marquis:

> . . . in an organizational setting where the owner of an organization or his representative can accurately evaluate the findings of a project in terms of organizational goals, he can encourage the researcher who shows high probability of solving such problems. Also he can reward the researcher in relation to the extent to which the researcher aids in problem solution. As a consequence the researcher is motivated to seek solutions to difficult but "relevant problems" in preference to less relevant but easier (in terms of the paradigm) problems. In seeking a solution to the difficult problems the researcher at times must abandon traditional methods and thinking.[42] [parentheses added]

These findings are reminiscent of Morton's account discussed above. In both cases the problem-framing process is crucial and the role of the research manager as a selecting "demon" is important. Gordon and Marquis summarize their findings by noting that the innovations they studied were most frequent when there was high visibility of the consequences of research, when there was good communication between the researcher and his administrative superior, and when the researcher was free to specify research procedures (i.e., the exercise his competence once goals were set).

The recognized importance of research to innovation and the expanded investment in it has raised interest in finding out more precisely the relationship between basic scientific knowledge and technological advance, and the extent to which basic research is a prerequisite for applied research and development. Not surprisingly, hard-headed managers and politicians are anxious to know what they are getting for their investment in R&D and particularly in basic research. Although the question is by no means settled, some findings are available. Marquis and Allen, after studying communications patterns in technology, reached conclusions that were consistent with those of Derek Price and others who found that technology develops somewhat independently of science and builds instead on prior technologies. However, this conclusion varies from field to field. While it may be true in mechanics, the role of science is far more important in such fields as electronics and atomic energy. However, except for those cases where "gap-filling science" is required—that is, where the realization of a technological opportunity requires new basic research—the science employed by the technologist generally is the science he learned while still a student.[43]

Recent studies conducted under the sponsorship of two U.S. government agencies have thrown additional light on these questions. Project HINDSIGHT, conducted by the Department of Defense, studied the critical R&D events leading to the development of 20 weapons systems since 1940. It found that "undirected" research played an insignificant role in the development of the weapons systems, while breakthrough events in technology were of overwhelming importance.[44] A study undertaken by the National Science Foundation, TRACES, comes to some rather different conclusions. Its findings, based on a study of 5 carefully selected technologies, demonstrate that basic research breakthroughs were essential for the later technological breakthroughs, although there may have been no realization of the practical benefits of the former at the time of discovery. The TRACES findings are not inconsistent with those of Marquis and Allen and Price in pointing out that the technological innovation was based on the science from a previous generation and involved an average lag of about 30 years.[45] Further, the TRACES findings were not necessarily inconsistent with those of the HINDSIGHT project. HINDSIGHT only traced R&D events back to 1940 and as a result would not have included many of the key basic science events that made later technologies possible. TRACES, on the other hand, found it necessary in some cases to go back to the nineteenth century.

If one assumes that basic research is necessary to "replenish the well," is there some way to overcome the generational time lag through changes in institutional arrangements and make basic research more relevant at an earlier date? It has been noted for instance that

> . . . it is probable that the 30-year cycle has often been due to the fact that the technological relevance of new R&D events has not been understood by, or not known to, the competent people or university.[46]

It seems likely that if the diffusion of new basic science knowledge is not left to chance, the rate of its utilization in new technology can be quickened.

In a useful review of a number of recent studies of the innovation process, Price and Bass summarize the relationship between research and innovation as follows:

1. Although the discovery of new knowledge is not the typical starting point for the innovative process, very frequently interaction with new knowledge or with persons actively engaged in scientific research is essential.
2. Innovation typically depends on information for which the requirement cannot be anticipated in definitive terms and therefore cannot be programmed in advance; instead, key information is often provided through unrelated research. The process is facilitated by a great deal of freedom and flexibility in communication across organizational, geographical, and disciplinary lines.
3. The function of basic research in the innovative process can often be

> described as meaningful dialogue between the scientific and technological communities. The entrepreneurs for the innovative process usually belong to the latter sector, while the persons intimately familiar with the necessary scientific understanding are often part of the former.[47]

Thus, while research certainly has a role in the innovation process, that process is typically non-linear and unpredictable. The prospects of planning or programming innovation therefore are unlikely, although it is felt that the creation of conditions for the free flow of information among scientists, technologists, and entrepreneurs (the "coupling" of the generators and users of information) is a worthy goal to be pursued.[48]

Research and Innovation in Low-Income Countries

The bulk of what has been said above has been drawn from experiences with research and innovation in the industrialized countries. There is no reason to think that the problems of the research and innovation processes are not universal and would not obtain in the low-income countries as well. There is, however, good reason to think that there are many other problems affecting research and innovation that are unique to the low-income countries and appear to be part of a syndrome of underdevelopment.[49]

An early characterization of this syndrome was offered by Steven Dedijer in an article appropriately entitled "Underdeveloped Science in Underdeveloped Countries." Dedijer argued that as part of the syndrome, scientists in low-income countries can be expected to feel isolated from the international scientific community and to become demoralized. Scientists are surrounded by an inhospitable institutional environment that neither understands nor sustains their work. Finally, scientists will find themselves in a cultural context that is not scientific or technological in its view of the world and its orientation to solving problems.[50]

Recalling that the great majority of experienced scientists in low-income countries received advanced training in the industrialized countries, we may assume for the moment the universality of the internal dynamics of science discussed above. It then becomes possible to explain some of the characteristics noted by Dedijer. The sense of isolation is particularly important. Often, scientists in low-income countries lack the physical and social means to participate in the international information-recognition exchange. Laboratories, equipment, and funds for research are generally inadequate and the close person-to-person contact with colleagues in the same field is either absent or inadequate. The scientist, inasmuch as he is cut off from the means to participate in the i-r exchange, is therefore cut off from the rewards that are most meaningful to him. Furthermore, he can be expected to feel considerable strain between his role as a trained scientist with its orientation to the international paradigm and his role as a privileged national intellectual faced with the technological innovation tasks of economic development that are often scientifically mundane.

The implications of an inhospitable institutional environment are legion. The scientist can be expected to face such problems as bureaucratic red tape, inadequate research support, low salary, younger scientists coming out of indigenous educational institutions who are incompetent to serve under him, the feeling that the utility of his contribution is not recognized, and so on. These problems may be encountered separately, but are more often a part of a syndrome.

Finally, the non-scientific cultural context can be expected to have a number of effects. First, it will limit the numbers of available technicians and support personnel needed to fill out the staff at research installations. It will be a constant reminder of the great gap existing between the Western-trained elites and the masses; between a modern culture and a culture presumed to be backward. The result can only be the worsening of the scientists' feelings of isolation and frustration. Perhaps most importantly, an absence of an awareness of and appreciation for the activities of R&D will severely limit the demand for the products of R&D. This lack of scientific consciousness among the general populace seriously constrains the impact that R&D will have on modernization.[51]

A more recent report on science in low-income countries made by a group from the University of Sussex is consistent with much that has been said above.[52] The Sussex group found four main factors limiting the impact of science and technology in the low-income countries. These included (1) the weakness of scientific institutions, their low level of performance and impact on the environment; (2) the problem of access to the science and technology of industrialized countries; (3) the obstacles to the application of new technology arising from underdevelopment itself; and (4) the "weight" and orientation of science and technology in the industrialized countries and the influence these have in the low-income countries.

This fourth point deserves further comment. In 1967, Oscar Varsovsky described the weight of advanced science and technology on low-income countries as "scientific colonialism." Scientific colonialism results in the compartmentalization and isolation—geographically and sociologically—of science; it distorts research priorities and finally results in the widespread demoralization of the scientific community.[53] The Sussex Group has taken these ideas a few steps further. They note first that the weight of international science is responsible for internal brain drains within the country. In order to enter into the international science game, there is the feeling that a certain critical mass of scientific talent must be concentrated in one place. The result is the creation of a research enclave with little impact on the national environment since it maintains an orientation to the international scientific community.

A second effect of international science is the often-discussed external brain drain, where indigenous scientists emigrate from their native countries to the metropolitan research centers of the world. Finally, the Sussex Group noted the effects of the composition of the stock of knowledge that characterizes advanced science and technology. Much of the attention of the international scientific community is focused on research problems that are simply irrelevant to the needs of low-income countries. A corollary is that many advanced technologies of the industrialized nations are ill-suited—economically and culturally—to be adopted and used in the non-industrialized nations. Even the technological breakthroughs in the former that appear to have no immediate relevance to the latter turn out, in many cases, to adversely affect the economic development of the low-income country. A good example of this is the revolution in the technology of synthetic materials. The synthetics can be produced more cheaply than the importation of the natural primary products that are among the chief exports of the low-income country.[54]

The growing concern for the impact of the international community on the scientific development of low-income countries is an encouraging sign. Although our knowledge of this impact remains embryonic, there can be little doubt that the economic, cultural, and at times political penetration of the science systems of low-income countries by the industrialized countries via indigenous, anti-developmental elites has distorted and in a real sense retarded scientific development.[55] This foreign penetration has been true even in China, which has been relatively isolated from the international system and which, as we shall see, has taken vigorous steps to control it.[56]

Social and cultural features of low-income countries take on additional significance when one speaks about innovation generally as opposed to research only. There appears to be a kind of vicious (and unfortunately downward) cycle of causes replacing effects and effects replacing causes. Knowledge of the techno-scientific societies of the industrialized nations lead to expectations that technological innovations will contribute to development, yet the conditions for innovation discussed above are rarely present in low-income countries. As discussed above, these included (1) a scientific and technological capability; (2) a market demand; and (3) an agent which transforms this capability into goods and services.[57]

At a slightly different level, the importance of communication between researchers and those concerned with production was noted. In the low-income countries the problems of communication are far more acute. To begin with, the physical means for communication are themselves underdeveloped, which makes overcoming spatial barriers far more difficult than in the industrialized nations. The psychological distance resulting from different subcultures and languages, for instance between a foreign-trained scientist or engineer and a local entrepreneur, is also more difficult to overcome. This problem of psychological distance is further exacerbated by wide status differentials between educated elites and those closest to production who are most likely to perceive innovative opportunities.

Finally, more often than not, the motivation to communicate is not there. The motivation of scientists and engineers working in government research institutes is often limited to communication with each other, or with their colleagues abroad, rather than with local producers. Both private and state producers, on the other hand, are often overly secretive about their operations in their fear that communication will endanger their competitive position. Where communication between the research and production sectors is rarely tried, the benefits are generally not appreciated.

The failure of communications between R&D people and producers can be thought of as a problem of social integration that is well known to students of modernization. The communication problem is certainly not insurmountable, but it does require extraordinary efforts (in comparison to the industrialized nations where much more communication can be taken for granted) to overcome.[58]

Communication, in turn, seems to be related to innovativeness in organizations and an entrepreneurial approach to management. These seem to be based on societal sanctions for such traits as "ego-riskiness," the legitimization of conflict, flexibility, a willingness to break orthodoxy and pose questions in a new way and seek new solutions–in short many of those traits that contribute to interpersonal competence and creative interpersonal interaction.

The use of interpersonal competence as it relates to innovation in modern organizations is *not* the same as the competence shown in interpersonal relations as defined by the norms of non-Western cultures. To say that a social system in a low-income country does not display the traits of interpersonal competence in Argyris' sense is not to say that it displays traits of incompetence in interpersonal behavior as measured by its own norms. Yet this distinction raises the profound question of the relationship between the two.[59] If the innovative traits discussed above must be present to successfully implement programs of research and innovation and if the indigenous cultures of many low-income countries do not display these traits, then must we infer that research and innovation programs are not likely to succeed? The logic of the argument so far suggests an affirmative answer.

However, the affirmative answer in no way implies an immutable situation. It merely suggests that a realistic science policy is one that incorporates the principle that "technological innovation is more than R&D." It suggests further that if governments are going to commit themselves to the development of an indigenous R&D capacity, their efforts will be wasted unless they coordinate these efforts with other efforts to change the environment in which R&D occurs. This conclusion, drawn from the argument developed in this chapter, is strikingly similar to the conclusion reached by the Government of India National Committee on Science and Technology in 1973. In its report, *An Approach to the Science and Technology Plan*, the Committee noted:

There are many reasons for the shortcomings and failures in the country's effort to mobilise effectively science and technology for economic and social development. But the chief among them are the deficiencies in the organisational structure of institutions for evolving science and technology policy and in implementation. It is most important to realise that the national Science & Technology Plan can succeed if, and *only* if, considerable managerial reform is brought about in our centres of scientific and technological activity and the administrative departments associated with them. Neither the alterations of organisational charts nor a rigid adherence to the dictum of 'first find a man and then build laboratories around him' are sufficient principles to work on today—*what is required are some basic reforms in the "working culture" of our scientific and administrative establishments. By working culture we mean the totality of administrative practices, organisational structures and the commitment, motivation and effectiveness of the people working in these organisations.* The scientific establishments in our country have, with rare exceptions, imbibed the culture of the administrative services, its extensive rules and regulations, its weighty statutes and precedents and its zealously guarded and regulated hierarchies and jurisdictions. Indeed, the two problems characteristic of this administrative system—hierarchy and detailed administrative regulatory mechanisms—are the very ones which are the most inimical to the development of effective scientific institutions. There is thus today a desperate need for innovations in both the style and methodology of administrative functioning inside and outside the scientific establishment if the latter are to fulfill the high expectations raised of the capability of science and technology for making major contributions to our development.[60] [italics added]

The logic of the above points to the need for a science policy that is concerned with micro-level organizational change directed towards the creation of innovative traits. It also points to one which is socially integrative in its support of a *network* of research, educational, and production units so arranged as to facilitate communication and maximize the possibility of unexpected discovery and innovation.

Perhaps most importantly, however, it highlights the question raised in the Introduction concerning the relationship between scientific development and societal change. It is often forgotten that the development of modern science in Europe in the seventeenth century occurred at a time of intense political, economic, and social change. This change in turn was linked to an ideology—the Puritan ethic—which both encouraged and legitimized it and which, as Merton has argued, legitimized the institutionalization of science as well.[61] Thus scientific development in the West was associated with the taproot of societal change in the West.

At a minimum, then, it would seem that developmental science policy must be concerned with societal change as a correlate to, or prerequisite for, scientific development. Where possible, it must be associated with whatever the particular taproot of social change might be in the society for which it is intended.

We have attempted here to piece together some scattered views on science, technology, and innovation from diverse fields of inquiry in order to gain a more comprehensive, and hopefully realistic, view of what science policy entails if it is to be authentically developmental. We have argued that innovation is not a simple function of R&D and that the establishment of a scientific tradition, which contributes in a major way to development in a low-income country, is not a simple function of state expenditures, the training of technical manpower, or the establishment of research organizations. Rather, the establishment of such a tradition is a function of a complex array of factors dealing with micro- and macro-levels of societal change, as well as with expenditures, R&D institutions, and R&D manpower. It is from the perspective developed here that we will consider the subject to which we now turn—Chinese efforts to foster science and innovation.

Notes

1. OECD, *The Conditions for Success in Technological Innovations* (Paris: OECD, 1970), p. 56.

2. OECD, *Science Growth and Society: A New Perspective* (Paris: OECD, 1971), p. 17.

3. Quoted in Ibid., p. 11.

4. Graham Jones, *The Role of Science and Technology in Developing Countries* (London: Oxford University Press, 1971), p. 15.

5. Norman W. Storer, *The Social System of Science* (New York: Hall, Rinehart and Winston, Inc., 1966), p. 3.

6. Ibid., p. 75. See also Warren O. Hagstrom, *The Scientific Community* (New York: Basic Books, 1965).

7. Marion J. Levy, *Modernization and the Structure of Society* (Princeton: Princeton University, 1966), p. 156.

8. Martin Landau, "Linkage, Coding, and Intermediary: A Strategy for Institution-Building," *Journal of Comparative Administration* Vol. 2, No. 4 (February 1971), p. 408. This excerpt is reprinted by permission of the Publisher, Sage Publications, Inc.

9. Jacques Ellul, *The Technological Society*, (trans. John Wilkinson), (New York: Alfred A. Knopf, 1964).

10. Cited in Albert H. Teich (ed.), *Technology and Man's Future* (New York: St. Martin's Press, 1972), p. XII.

11. Thomas S. Kuhn, *The Structure of Scientific Revolutions* (Chicago: The University of Chicago Press, 1962), p. 10.

12. Derek J. De Solla Price, "Is Technology Historically Independent of

Science? A Study in Statistical Historiography," *Technology and Culture* 6 (1965).

13. Cited in Edwin Mansfield, *Technological Change* (New York: W.W. Norton and Co., Inc., 1971), pp. 42-43.

14. OECD, *The Conditions for Success in Technological Innovations*, p. 19.

15. Ibid.

16. Ibid.

17. See, for instance, Warren O. Hagstrom, *The Scientific Community* (New York: Basic Books, 1965).

18. J.M. Ziman, *Public Knowledge* (Cambridge: Cambridge University Press, 1968), p. 24.

19. William Kornhauser, *Scientists in Industry* (Berkeley: The University of California Press, 1962), p. 1.

20. Ziman, *Public Knowledge*, p. 24.

21. Hagstrom, *The Scientific Community*, p. 150.

22. Donald G. Marquis and Thomas J. Allen, "Communications Patterns in Applied Technology," *American Psychologist* Vol. 21, No. 11 (November 1966), p. 105.

23. Price, "Is Technology Historically Independent of Science?" p. 501.

24. Louis Morton, "War, Science and Social Change," in Kalman H. Silvert (ed.), *The Social Reality of Scientific Myth* (New York: American Universities Field Staff, Inc., 1969), p. 25. Reprinted by permission of the publisher.

25. Elting Morison, *Men, Machines and Modern Times* (Cambridge: MIT Press, 1966).

26. Victor A. Thompson, *Bureaucracy and Innovation* (University, Alabama: The University of Alabama Press, 1968), pp. 16 and 18.

27. Ibid., p. 47.

28. J.A. Morton, *Organizing for Innovation* (New York: McGraw-Hill Book Co., 1971), p. 3.

29. Ibid., p. 19.

30. OECD, *The Conditions for Success in Technological Innovation*, p. 131.

31. See Morton, *Organizing for Innovation*, p. 61ff. Also, Royden C. Sanders, Jr., "Interface Problems Between Scientists and Others in Technically Oriented Companies," in Karl B. Hill (ed.), *The Management of Scientists* (Boston: Beacon Press, 1964).

32. Morton, *Organizing for Innovation*, pp. 64-65.

33. Chris Argyris, *Organization and Innovation* (Homewood, Illinois: The Dorsey Press, 1965), pp. 43 and 203-204.

34. See Ibid., p. 5ff.

35. OECD, *The Conditions for Success in Technological Innovation*, p. 11.

36. Ibid., p. 13.

37. Charles A. Dailey, *Entrepreneurial Management* (New York: McGraw-Hill Book Co., 1971), p. 21.

38. Morton, *Organizing for Innovation*, p. 99.

39. Kornhauser, *Scientists in Industry*.

40. Howard Vollmer, et al., *Adaptations of Scientists in Five Organizations: A Comparative Analysis* (Menlo Park, California: Stanford Research Institute, 1964).

41. Leo Meltzer, "Scientific Productivity in Organizational Settings," *Journal of Social Issues* 12 (1956), pp. 32-40.

42. Gerald Gordon and Sue Marquis, "Freedom, Visibility of Consequences and Scientific Innovation," *American Journal of Sociology* Vol. 72, No. 2 (September 1966), p. 199. © by The University of Chicago. All rights reserved.

43. Marquis and Allen, "Communications Patterns in Applied Technology." See also Price, "Is Technology Historically Independent of Science?"

44. See discussion in OECD, *The Conditions for Success in Technological Innovation*, p. 82ff.

45. Illinois Institute of Technology Research Institute, *Technology in Retrospect and Critical Events in Science* (Washington: National Science Foundation, 1968).

46. OECD, *The Conditions for Success in Technological Innovation*, p. 87.

47. William J. Price and Lawrence W. Bass. "Scientific Research and the Innovative Process," *Science* 164 (May 16, 1969), p. 804. Copyright 1969 by the American Association for the Advancement of Science.

48. Ibid., pp. 802-803, 805.

49. For a good recent discussion of these from a managerial perspective and a plea for more entrepreneurial managers for technical projects in low-income countries, see Board on Science and Technology for International Development, Office of the Foreign Secretary, *Research Management and Technical Entrepreneurship* (Washington: National Academy of Sciences, 1973).

50. Steven Dedijer, "Underdeveloped Science in Underdeveloped Countries," *Minerva* Vol. 2, No. 1 (August 1963), pp. 61-81.

51. See a discussion of this problem based on conversations with Asian science administrations by C.H.G. Oldham, *Saturday Review* (March 6, 1965).

52. "Science in Underdeveloped Countries." Statement made by the Sussex Group to the Advisory Committee on the Application of Science and Technology to Development, *Minerva* Vol. 9, No. 1, (January 1971).

53. Oscar Varsovsky, "Scientific Colonialism in the Hard Sciences," *American Behavioral Scientist* Vol. X (June 1967), pp. 22-23.

54. See Sussex Group, "Science in Underdeveloped Countries," pp. 105-107.

55. A most convincing case for this thesis, based on Latin American experience, is made in Amilcar Herrera, "Social Determinants of Science Policy in Latin America," *Journal of Development Studies* Vol. 9, No. 1 (October 1972), pp. 19-37.

56. For a discussion of Soviet influence on Chinese science during the 1950s see Richard P. Suttmeier, *The Chinese Academy of Sciences: Institution*

Building in a Research Oriented Organization, unpublished doctoral dissertation (Department of Government, Indiana University, 1969), pp. 70-91.

57. OECD, *The Conditions for Success in Technological Innovation*, p. 11.

58. This extraordinary effort has characterized the activities of the Korean Institute for Science and Technology, and the Central Leather Research Institute in Madras, India, two of Asia's more successful R&D establishments. For a discussion of the latter, see Y. Nayudamma, "Promoting the Industrial Application of Research in an Underdeveloped Country," *Minerva* Vol. V, No. 3 (Spring 1967), pp. 323-339.

59. An explicit recognition of this problem in the Indian case and a creative solution that appears to be consistent with the argument developed in this chapter is found in Y. Nayudamma, "Decentralised Management of R&D in a Developing Country," *Minerva* Vol. 11, No. 4 (October 1973), pp. 516-536.

60. Government of India National Committee on Science and Technology, Department of Science and Technology, *An Approach to the Science and Technology Plan* (January 1973), quoted in Board on Science and Technology for International Development, Office of the Foreign Secretary, *Research Management and Technical Entrepreneurship*, p. viii.

61. Robert K. Merton, *Science, Technology and Society in 17th Century England* (New York: Howard Fertig, 1970).

2 Policy Shifts and Abiding Principles for Chinese Scientific Development

Class struggle, the struggle for production, and the struggle for scientific experiment are the three great revolutionary movements for building a mighty socialist country.

—Mao Tse-tung

As the preceding chapter has shown, science policy and the unfolding of research and innovation programs do not exist independently. Rather, both policy and science-related activities are linked by complex interdependencies to other policies and to social institutions in the environment. While science policy is intended to shape research and innovation activities in the interest of societal change, particular science policies are nevertheless products of socio-political and economic forces. One way of expressing these complex interdependencies is to invoke the well-known metaphor of a cybernetic system. According to the metaphor, economic and socio-political factors are seen as inputs that are processed by the political system. Policies are seen as outputs directed at effecting the achievement of some social goal. An effectively working system will include a feedback loop, which contains messages concerning the efficacy of the policies, and which provides the information necessary for the system to stabilize itself.

The Phenomenon of Policy Shifts

On the face of it, the metaphor of the self-stabilizing, equilibrium-model social system would seem highly inappropriate for representing events in China. The history of the People's Republic is full of major, and often sudden, destabilizing policy changes and variations in the political environment in which policy is made. These changes do, however, seem to show certain patterns. Skinner and Winckler, for instance, in studying policy changes as they affect rural China have noted a periodicity in the changes.[1] Some periods seem to have characteristics that recur in a later period, and this pattern of recurrence has been sufficiently regular to lead them to speak in terms of a "policy cycle."

In gross terms, the cycle consists of a "liberal" period, a "radical" period, a period of retrenchment, and then another liberal period. Although the cyclical thesis has been advanced convincingly, the image of the cycle does have one major drawback, at least as far as science policy is concerned. The essential flatness of a cycle does not permit an expression of progress or development. Thus, there is no way to incorporate an element of social learning, which it will be argued, is an essential feature in shifts in phases.[2] A more appropriate image, therefore, might be that of a spiral.

Other students of contemporary China have simply noted the shifts from "liberal" to "radical" policies without invoking the cyclical image.[3] From the Chinese point of view, as reconstructed by Schurmann, the shifts in policy and politics are related to the Maoist notion of contradictions and a dialectical conception of Chinese society.[4] In a recent expression of this periodicity, Chinese leaders are reported to expect radical upheavals in Chinese society every seven or eight years.[5]

Whether one chooses cycles, spirals, or dialectics, three observations about policy shifts in China are in order:

1. The shifts are usually not in one policy area only. They pertain instead to clusters of policies and, as Skinner and Winckler have noted, to constellations of goals, uses of power, and patterns of compliance. They are, in short, systemic in their implications.
2. There does indeed seem to be a process of social learning that accompanies the shifts. Thus, a shift from a radical phase to one of retrenchment or liberalization does not necessarily mean that valuable experiences, or mistakes, from the radical period are ignored. Instead, useful social innovations of previous periods persist.
3. The process of shifting policies may be as important for China's development as the specific content of the policies. The twenty-odd years of the history of the People's Republic points to the institutionalization of societal-wide experimentation and a certain flexibility in avoiding such developmental *culs de sac* as the institutionalization of administrative patterns that prematurely stabilize situations requiring further change.

The idea of societal experimentation as practiced in China can be represented analytically. At any given point in time, the social system is said to be in state m_{t_1} (see Figure 2-1). A "state" of the social system is understood in terms of the organizational structure of the system, the goals that organization embodies, and the policies adopted to meet the goals. The social system in state m_{t_1} embodies the intermediate goals a and c, the realization of which will move the system to state m_{t_2} and in the direction of end goals A,B,C,D. Societal experimentation is begun when there is dissatisfaction among at least some of the system's actors with state m_{t_2}. The achievement of goals a and c is a necessary but not sufficient condition for moving the system towards end goals, A,B,C,D. In the language of the cybernetic system metaphor, demands for the fuller realization of goals b and d create an input overload for the system in state m_{t_2}.

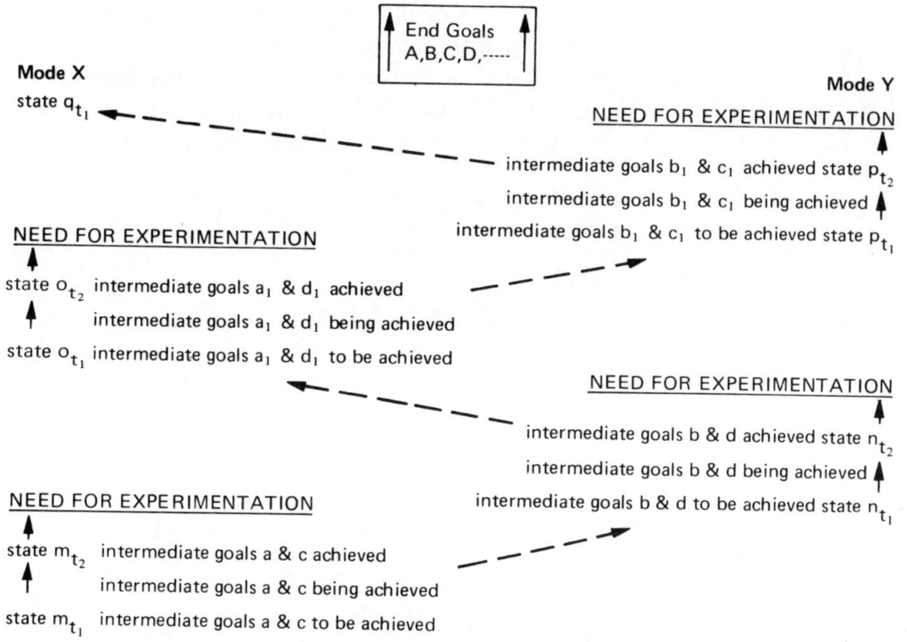

Figure 2-1. State Changes for the Progressive Achievement of Intermediate Goals.

Societal experimentation presumes a vision of an alternative mode of the system, state n_{t_1}, which embodies goals b and d and which unlike state m_{t_2}, is structurally suitable for achieving these goals. When the input demands (which are made in the interest of achieving goals A,B,C,D) reach a certain threshold, and an alternative vision is available that would move the system towards the end goals by the realization of intermediate goals b and d, there is a shift in the system from state m_{t_2} to state n_{t_1}. However, the intermediate goals embodied in state n_{t_1}, even when realized at state n_{t_2}, are again necessary but not sufficient. Since the intermediate goals exist in a series, a,b,c,d; a_1,b_1,c_1,d_1; a_2,b_2,c_2,d_2;–, and since states m_{t_2} and n_{t_2} only realized goals a,c and b,d, there is again pressure to change the state of the system from state n_{t_2} to a state that will move the system towards the next set of intermediate goals, perhaps a_1 and d_1.

In short, the system in one mode (X) allows for states that are structurally suitable for the achievement of certain kinds of goals but which are unsuitable for achieving others. The system in the other mode (Y) allows for states that are suitable for achieving different goals from those achievable under mode X (see Table 1-1). Thus, a situation of dynamic tension is suggested. All states of the system are directed towards end goals (which to be sure are not immutable), but no one state can satisfactorily achieve a given set of intermediate goals.

This rather complex analytic representation becomes considerably more complex when one "interprets" it with data on Chinese development. First, we may note states *m,o*, and *q* are classified as states of "regularization," and correspond to the 1949-1957, 1961-1965 and 1972-present periods (see Figure 2-2). States *n* and *p* are classified as states of mobilization and correspond to the

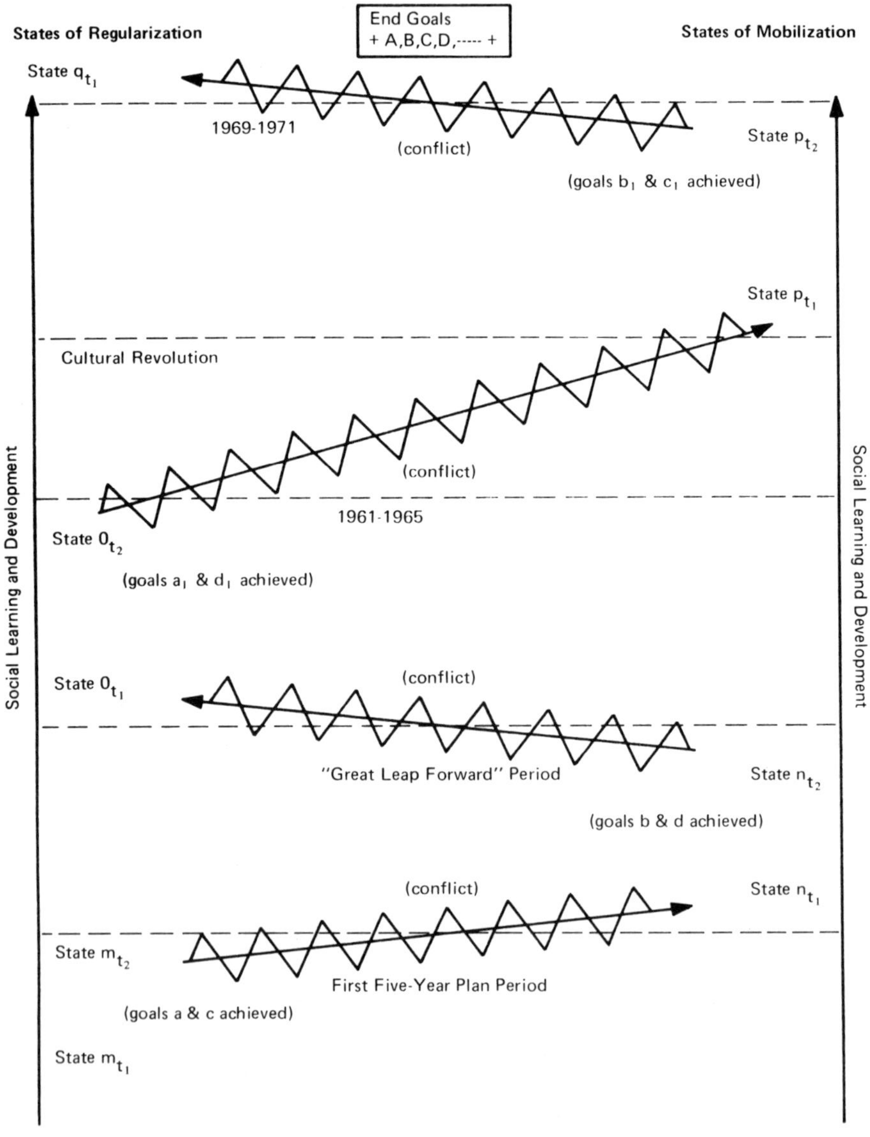

Figure 2-2. State Changes and Policy Shifts in China.

1957-1960 and 1969-1971 periods. The transitions from one state to another are represented by a jagged conflict line. This is meant to show that the policy shifts intended to move the system from one state to another are accompanied by political conflict. As we know, each significant policy change has been accompanied by a gain in power by some Chinese actors at the expense of others precisely because over time, different actors come to have vested interests in different states. The period of the Cultural Revolution is represented by a broadened conflict line to indicate that the change from state *o* to state *p* was marked by a period of conflict of unusual intensity and length. Indeed, the level of conflict called into question the integrity of the political system.

An important step in extending the metaphor advanced here is to identify a compass-like mechanism that assists the actors in the social system in deciding when and how to change states without losing the course to the achievement of the end goals. Such a mechanism would function like a standard, a template, or perhaps a map of ultimate objectives against which interim means-ends formulae could be measured. Thus, one may think of the transformation of inputs into outputs as setting interim ends and of devising means to achieve those ends. The outcome of the application of a means-ends formula is reported back to the system via the feedback loop and is then measured against the system's standard for means and ultimate ends.

This standard in contemporary China is an amalgamation of ideology and the accretion of developmental experience we might call social wisdom. Ideology has played a standard- or map-like role for the Chinese and has undoubtedly contributed an additional measure of social integration that has made possible the evolution of a system capable of drastic experimentation without experiencing total breakdown.[6] However, the actual influence of ideology in policy has varied over time, and the interpretation of ideology for policy has been a function of the quantity and kind of social wisdom with which it has been amalgamated.

The above, somewhat abstract, discussion is intended to provide a simplified overview of the dynamics of Chinese development. The striking aspect of that development has been the major policy shifts that have occurred at least four times in a little over twenty years. It becomes an intriguing problem to explain why they occurred and to assess their significance.

This study attempts to do this for science policy and its implementation. It is based on the assumption of considerable similarities in the policies and pronouncements of the 1949-1957 and 1961-1965 periods on one hand (which we have labeled periods of regularization), and the 1957-1960 and 1969-1971 periods on the other (which we have labeled periods of "mobilization"). Before identifying and analyzing the dynamics of policy shifts, it is necessary to review the end goals relating to science and technology China has set as well as the ideological standard against which performance is gauged.

Goals and Ideology for Science

The goals for science and technology in the new China are in a general way part of the heritage of modern China's hundred-year struggle with the problems of modernization. Throughout this struggle the development of both science and technology was to be realized as a goal. However, in a wider framework, the achievement of that goal was for essentially utilitarian purposes. Such a utilitarian orientation towards technology is expected, but science too has been consistently linked with non-scientific purpose. The idea of "science for the sake of science" has simply not been a significant cultural theme in China during the last one hundred years.[7] Instead, both science and technology were to contribute to the achievement of such national goals as the military and economic strengthening of the nation, and by the 1920s, to the abolition of Confucianism and the creation of a modern scientific culture.

To some of China's non-Marxist enthusiasts for scientific culture, the most promising path for realizing that culture lay in the wholesale emulation of the West. But, an alternative to the Westernization path to cultural transformation was also gradually being articulated within the Communist movement and particularly within the writings of Mao Tse-tung. Thus, in the final section of *On New Democracy* (1940), Mao set forth criteria for "a national, scientific and mass culture." Because of the relevance of these words for later developments in the People's Republic, it is appropriate to cite them selectively at length here.[8]

Mao notes that the new-democratic culture will belong "to our own nation" and bear "our national characteristics."

> To nourish her own culture China needs to assimilate a good deal of foreign progressive culture, not enough of which was done in the past. We should assimilate whatever is useful to us today not only from the present-day socialist and new-democratic cultures but also from the earlier cultures of other nations, for example, from the culture of the various capitalist countries in the Age of Enlightenment.

However,

> To advocate "wholesale Westernization" is wrong. China has suffered a great deal from the mechanical absorption of foreign material. . . .

Mao went on to note,

> New-democratic culture is scientific. Opposed as it is to all feudal and superstitious ideas, it stands for seeking truth from facts, for objective truth and for the unity of theory and practice. On this point, the possibility exists of a united front against imperialism, feudalism and superstition between the scientific thought of the Chinese proletariat and those Chinese bourgeois materialists and natural scientists who are progressive, but in no case is there a possibility of a united front with any reactionary idealism. . . .

A splendid old culture was created during the long period of Chinese feudal society. To study the development of this old culture, to reject its feudal dross and assimilate its democratic essence is a necessary condition for developing our new national culture and increasing our national self-confidence, but we should never swallow anything and everything uncritically. It is imperative to separate the fine old culture of the people which had a more or less democratic and revolutionary character from all the decadence of the old feudal ruling class. . . .
New-democratic culture belongs to the broad masses and is therefore democratic. It should serve the toiling masses of workers and peasants who make up more than 90 percent of the nation's population and should gradually become their very own. . . .

Mao also saw in science a tool for combatting oppression imposed by nature.

For the purpose of attaining freedom in the world of nature, man must use natural science to understand, conquer, and change nature and thus attain freedom from nature.[9]

Thus, for Mao, science would be incorporated into the new society as a means of liberation from both nature and traditional culture. At the same time, however, the new culture, and implicitly science, should conform to the criteria of being *national* and *democratic*.

The most formal statement of Chinese goals for science and technology in the early years of the People's Republic is found in the Common Program of the Chinese People's Political Consultative Congress, issued in 1950:

Efforts shall be made to develop the natural sciences in order to serve construction of industry, agriculture, and national defense. Rewards should be given for scientific discoveries and inventions. Dissemination of scientific knowledge should be strengthened.

In addition, the regime declared "love of science" to be a basic national virtue.[10] Thus, at the outset of the People's Republic, we see the persistence of long-standing utilitarian interest in, and goals for, science and technology. To summarize, these are: (1) to serve economic construction; (2) to serve national defense; and (3) to serve cultural transformation through the widespread dissemination and popularization of science and technology.

Chinese ideology contains not only goals for science but also a reasonably well-articulated doctrine of science and society. It is this doctrine that serves as part of the performance standard mentioned above. It is exceedingly important to recognize the existence of this doctrine in China. Not all countries committed to a course of rapid development have it, except to the extent that an idealist interpretation of the evolution of science-society relations in the West is held implicitly by Western-educated political and scientific elites. It is, therefore, necessary to explicate this doctrine before proceeding to a discussion of the evolution of science *policy*.

Perhaps the most comprehensive and authoritative statement about science-society relations in China is a speech by Ch'en Po-ta to the Chinese Academy of Sciences (CAS) in July 1952. This speech can be taken as the basis of our discussion.[11]

Ch'en's speech came at the end of the ideological reform movement within the Academy that had begun in December 1951.[12] The speech appears to be in response to the scientists' uncertainty about their roles, which was a result of the campaign. The speech had three parts. The first dealt with the relationship between scientists and Chinese Communist Party cadres as well as the relationship between older scientists and younger ones. The second concerned the orientation of the work of CAS, and the third was concerned with what Ch'en called "scientists and patriotism."

The first section was in response to conflicts between Party cadres and the non-Party scientists. Underlying this operational issue, however, was a more fundamental one that had also appeared in the Soviet Union in its early days. The conflict, which concerned the trustworthiness of "bourgeois" scientists and their roles in Chinese modernization, was one manifestation of the larger problem of fostering "Red experts." We see in Ch'en's speech both the Leninist formula for dealing with this issue and what can be considered an emerging Maoist formula. Ch'en quotes Lenin:

> We have convinced Russia, we have won Russia for the toilers from the exploiters, we have crushed the exploiters. Now we must learn to administer Russia. And to do that we must learn to be modest and to respect the efficient work of the "experts in science and technology."[13]

The message to the Party cadres seems clear: put a measure of trust in the scientists, be prepared to learn from them, assist them, and don't arbitrarily interfere with their work. Such would be the operational or policy directive, but the ideological assumption behind the directive would be that since Marxism-Leninism is a scientific doctrine, scientists should be left to accept it on their own, through their research on the materialist basis of nature.

According to Ch'en, Lenin felt a scientist or engineer will "come to accept communism through the data of his science and in his own way."[14] While it was certainly necessary for scientists to become "Red"—for only then would science truly serve the people—scientists were to reach ideological conversion in their own way. By this Leninist standard, there would be no need for an ideological remolding campaign. But indeed, the scientists had just been through one. The thrust of the campaign to bring the scientists closer to the people and the fact that it occurred at all—indicating that the CCP was skeptical about the autonomous transformation of the scientists—marked two important aspects of an emerging Maoist doctrine of science. Ch'en writes:

> . . . Comrade Mao Tse-tung also told us long ago that intellectuals and scientists must associate themselves with the people if their work is to be crowned with

success, otherwise, they will be unable to accomplish anything. Experience of the last one hundred years teaches us precisely that only by joining the masses of the people and by cooperating with the Communist Party, will scientists be able to find their future.[15]

The second section of Ch'en's speech concerned the question of "the orientation of the Academy's work."

The large amount of work by the Academy should meet the urgent demands of the people and serve the immediate task of the State, and the planning of national construction.[16]

Thus, science was to serve production but could only do so when theory and practice were united. The unity of theory and practice would be attainable only when science was associated with the masses.

The close relationship of science, production, and politics that was the essence of the Party's doctrine of science was manifested in the historical explanation for the failure of modern science to develop in China. Ch'en noted that in the years prior to 1949, no great scientific achievement came from China in spite of the fact that at least in the last thirty years, there had been some scientists working conscientiously on research. Ch'en asked whether this was due to a unique Chinese incompetence in science.

Decidedly not. It is because under reactionary rule and imperialist domination, it was impossible to develop our industries and to meet the needs of the people.[17]

In Ch'en's view, then, those who took up scientific studies had not been given encouragement or material support; hence, the seeds of science had not developed.

Accordingly, the development of science in the West would be considered by Ch'en a function of industrialization, which in turn could be explained by the historical laws of Marxist science. Science had not developed in China because the socio-economic conditions had not been right. While Europe was industrializing, China was, to the Party, still under feudalism. Even when the feudal past was abandoned, China remained under "reactionary rule and imperialist domination" until 1949.

This (Marxist) sociological—or *externalist*[18]—intepretation of the development of science has psychological value as well as expected policy consequences. For the externalist, scientific backwardness can be explained by reference to the evils of the old society—a social organization that would continue to frustrate science. Needham has argued that an *internalist* explanation of the rise of modern science only in Europe is ultimately postulated either on chance occurrence or on a disguised (or undisguised) statement of European racial and philosophical superiority.[19] Clearly, as Marxists and as Chinese, the members of the Chinese Communist Party could not accept either of these alternatives.

Feelings of inferiority in the face of European scientific accomplishments apparently has been a real problem in China and, as we saw in Chapter 1, in other low-income countries as well. It has been alluded to many times over the years by the Chinese media. As we shall see however, the Party, in keeping with its societal change strategies, has attempted to turn passiveness stemming from feelings of inferiority, into activism. This has been done by encouraging "struggle" against foreign and "bourgeois" scientific and technical "authorities," who represent the forces retarding China's social and scientific development.

Ch'en went on to say that in the past, scientists had been driven to *individual, private* research because neither appreciation nor support had been forthcoming from the society. In the new society, however, any tendency of a scientist to shut himself and his research off from society, would be a mistake since it would divorce him "from the most fundamental realities"—namely the lives of the masses and the realities of production. In the new society, science was to be "a common enterprise of the people as well as a common enterprise of the scientists"[20]—that is, science would no longer be an individual activity, as a method of collective research is gradually established.[21] Such collective effort would facilitate cooperation and the sharing of research findings. Collective, cooperative research would be amenable to *planning,*[22] which would improve efficiency and coordination in research and would reduce duplication of effort. Furthermore, science plans could be coordinated with economic plans for national construction.

Apart from the importance of common collectivized effort, Ch'en set three other conditions for the Academy's successful work. The first of these was to achieve the proper relationship between theory and practice. Following Mao's writings, neither theory nor practice would be neglected. "Some people suspect that in speaking of connecting theory with practice, we only want practice without theory. This is not so."[23] Citing certain areas of mathematics and astronomy as fields with no immediate practical value, Ch'en justified theoretical research in these fields by arguing that discoveries there may further other sciences with more obvious practical payoffs.

Ch'en's second condition concerned the selection of topics for research. In the new society, research was to be oriented toward the broad problems of economic construction and national defense.

> . . . What we ask for is that scientific work should contribute to the needs of the people and to national construction. What scientists have to pursue should, therefore, be chiefly what the people need in great quantities. . . . Research work must serve definite purposes and lead to the solution of important problems of a mass character.[24]

This condition would entail giving up narrow, trivial topics. Ch'en wrote that

> I have heard that some scientists want to reach and exceed the world level on certain trifling items and to accomplish something "sensational" on certain odd, narrow little topics.[25]

Thus, it was not for the Party or the Academy to simply prevent scientists from undertaking such work in a rude manner. Rather, efforts should be made to convince these scientists to concentrate on more socially significant topics. But if they insisted, they should be allowed to try anyway.

When they come up against snags and fail to obtain any result, they will come to understand that our opinion is right and they will correct their mistakes.[26]

It is important to realize the basis for Ch'en's certainty that socially insignificant research was bound to fail. Apart from the organizational means available to frustrate such research—assuming for the moment that Ch'en was insincere and that such measures as the withholding of funds could be used against these scientists—there were still ideological grounds for believing "trivial" topics would lead to failure. True knowledge, for Ch'en, only came from the application of "theory" in productive practice, following the Marxist-Leninist-Maoist "practice-theory-practice" principle of epistemology.[27]

At this stage of the development of CAS, the selection of appropriate topics was quite voluntaristic. Scientists were not simply asked to give up personal interests; rather they were to try to find areas where personal research interests coincided with societal interests. Here again, if scientists possessed the correct ideological orientation, the selection of topics would prove to be no problem. A true Red expert's interests would naturally be those of the people. But since most of the scientists were not Red experts, efforts would have to be made to show them the importance of working for the people. In the final analysis, however, and as a portent of things to come, Ch'en noted that "What we ask for is that scientists should be in a position to account for their research plans and the aims of their research."[28] This statement points to the important social control function of the planning system that was soon to be instituted.

The final condition set forth by Ch'en for the successful operation of CAS pertains to the work style of the scientist. Scientists were expected to work with diligence, simplicity, and frugality to accomplish successful research. They were to examine their work in light of societal goals, engage in self-criticism, and be willing to correct mistakes. The scientists and science administrators were to resolutely oppose a style of work where achievement at an early age would lead to conceit, self-satisfaction, and laziness.

In the final section of Ch'en's speech, he addressed himself to those problems of scientific development resulting from China's recent historical experience. It was necessary, according to Ch'en, to overcome feelings of national and cultural uncertainty, the bitter product of one hundred years of Chinese weakness. Chinese should show pride in their inherent scientific capabilities.

We must have self-confidence in making creations. Foreign imperialist aggression and domestic counterrevolutionary rule dragged China into a dark abyss. Consequently, there was a lack of self-confidence on the part of a certain number of people who thought that there was no hope whatsoever for things

Chinese. Some intellectuals of the comprador type even said that "the moon over the United States of America is brighter." People of this type can also be found in the midst of the scientists. They think that everything in the U.S.A. and Britain is good while nothing in China is good.[29]

One reason for the Chinese sense of inferiority in science, according to Ch'en, was the comparative youthfulness of science—its rather short life in China. But this—like the lack of experience of younger scientists—was not completely a bad thing:

Youth has its advantages *since it is not tied to old traditions* and is more ready to accept new things. In science, we can find unexplored virgin soil everywhere. *Anyone* who is willing to make serious efforts and devote his energy will meet with success in science, and the objective conditions for achieving success do exist.[30] [italics added]

The recognition of the need for these objective conditions to exist was viewed as a major step forward in the ideological progress of scientists. It marked the outgrowing of a naive vision once held by intellectuals that China could be saved by science and education without fundamental political and economic change—that is, without overthrowing the rule of feudalism and foreign imperialism.

Many well-intentioned educators, scientists and students who have stayed away from politics and dreamed that they could serve their country with their learning have equally been disillusioned. This is a good sign, for the awakening from such childish dreams marks the beginning of a prosperous and strong China.[31]

While Ch'en Po-ta's speech of July 1952 is an important statement of CCP policy towards the Academy in the years prior to serious efforts of the Party to lead CAS in conjunction with the First Five-Year Plan, it also contains most of the major elements that seemed to make up the Party's more general doctrine for science. These include, first, a radically sociological view of scientific progress derived from a fully externalist historiography of science. In its most radical form, this extreme sociological view postulates that *all* that is needed for the development of science is (1) the proper environment for science, which is naturally furnished by a socialist state, (2) the proper organization of research, and (3) the proper ideology of the scientists. Furthermore, this sociological approach yields a view of science as an organized, collective activity, the practical benefits of which can be maximized through planning. This sociological view entails the belief that research activity is governed by *laws* that are knowable.

A second important element in the CCP view of science contained in Ch'en's speech would be the principle of the unity of theory and practice. CCP understanding of scientific theory in Western philosophical language might be called "simple inductionism." Theory is regarded as merely the generalization of the sense perceptions of experience. Since the most important kind of experience for the CCP is the productive experience of the peasant and worker, the CCP had a ready ideological foundation for the policy of sending scientists to farms and factories to "learn from the masses," a policy adopted during mobilization periods. The intellectuals were to experience the practice of production and then were to "sum up" this experience—that is, to generalize "scientific" theory. While the theory-practice principle very definitely has policy consequences, the principle is not so unambiguous as to justify only one interpretation in policy. Hence, the tolerance for theoretical work in science has varied over time.

Two other ideas are closely related to the theory-practice principle. The first is that science must serve production. While this view conveniently relates to Chinese modernization policies, it also must be seen as an inference from the basic tenets of CCP ideology. Specifically, it too derives from the theory-practice account of knowledge. Theory is not justified until it is applied in the practice of production. Since the practice of production is the realm of the masses, an extension of the theory-practice principle gives rise to another feature of CCP views of science, which might be called the concept of "mass science." According to this concept, scientific *discovery* and scientific *verification* are ultimately derived from the masses. Discovery, as just noted, comes from summing up mass experience. With "mass verification," the final justification of a scientific theory lies not with the experiments of experienced scientists, nor with recognized scientific authority, nor with any other test of theory, but rather in mass experience. One of the implications of this principle is that much of science is to be reduced to the level of popular science. In Ch'en's speech we find but a hint of this principle.[32] However, beginning with the first mobilization period (1957-1960), this principle was considerably elaborated and programs to encourage worker and peasant "scientists" became widespread.

Another important element of CCP's views of science that is prominent in Ch'en's speech would be the element of cadre-scientist relations. These relations are part of the more general principle of Party leadership in science. Like theory-practice, "Party leadership" is not without its ambiguities; hence, the *policy* stemming from the principle has been subject to change, as have views of the precise nature of that leadership. One underlying assumption about science, however, has remained unchanged. Science has been regarded as analogous to a tool or a weapon that is used by one class or another. To insure that science is used properly as the instrument of the masses, science must be *led* down the "socialist road"[33]—as opposed to the "capitalist road"—by the CCP, the Party of the masses.

The question of the development of an indigenous Chinese scientific tradition can be viewed an additional element of the CCP views of science that is contained in Ch'en's speech. This tradition can be begun with the older generation of scientists, but the real hope of its establishment lies with the new generation. The new generation will not have been exposed to the extensive bourgeois influence felt by the older men who had studied abroad. Instead, it will be authentically both Red and expert. As we shall see this question of the differences between scientific generations has been a persistent theme in Chinese scientific development.

Finally, Ch'en's speech throughout is an argument for the primacy of politics in development. The criticism of those intellectuals who would save the nation through science or education can be interpreted as a more fundamental criticism of an essentially technical approach to development, one that focuses on technique as opposed to politics.

Ch'en's speech and other materials from the period indicate that in the early 1950s, the CCP had a fair idea of what was expected of science and technology in the new China.[34] Many of the tenets of the Party science doctrine were sufficiently vague, however, to permit varying interpretations when operational policy was devised, and when the organizational instrumentalities for implementation were established. Indeed, when the vagueness of doctrine failed to provide direction for policy (a condition which became worse as the process of development increased the complexity of the society), serious problems developed for policy making as our discussion of the events of the early 1960s will indicate.

In this chapter, we have proposed a schematic representation of the course of Chinese development. The schema calls for a mechanism that will both allow for flexible shifts in policy, while at the same time keeping the system oriented towards end goals. We have stated that ideology performs this function in China, and we have described selected ideological doctrines that are most relevant to scientific development. The next two chapters describe the evolution of policies and organizations for research and innovation within this general framework. Through an examination of that evolution, we shall see that both the goal-displacement effects of formal organizations and the means-ends calculus of the "social system of science" pose major challenges to the movement of the Chinese social system towards the realization of its end goals.

Notes

1. G. William Skinner and Edwin A. Winckler, "Compliance Succession in Rural Communist China: A Cyclical Theory," in Amitai Etzioni (ed.), *A Sociological Reader on Complex Organizations* (New York: Holt, Rinehart and Winston, Inc., 1969).

2. The possibility of a learning process is, however, noted by Alexander Eckstein in his cyclical interpretation of Chinese economic development. See his "Economic Fluctuations in Communist China's Domestic Development," in Ping-ti Ho and Tang Tsou (eds.), *China in Crisis* Vol. 1, Book 2 (Chicago: University of Chicago Press, 1968), pp. 671-728.

3. See for instance, Shinkichi Etō, "Communist China: Moderation and Radicalism in the Chinese Revolution," in James B. Crowley (ed.), *Modern East Asia: Essays in Interpretation* (New York: Harcourt, Brace and World, Inc., 1970), pp. 337-373; and Jan S. Prybyla, *The Political Economy of Communist China* (Scranton, Pa.: International Textbook Co., 1970).

4. Franz Schurmann, *Ideology and Organization in Communist China* (Berkeley: University of California Press, 1970), pp. 73-104.

5. Leo Goodstadt and Cheng Huan, "China: Purloined Letter," *Far Eastern Economic Review* 78 No. 49 (December 1973), p. 10. See also *China News Analysis* No. 951 (March 1, 1974) for a discussion of Mao's thinking concerning alternating periods of relaxation and revolutionary tensions in China's development.

6. For this functional interpretation of ideology, see Schurmann, op. cit., Chapter I; and, more generally, Clifford Geertz, "Ideology as a Culture System," in David Apter (ed.), *Ideology and Discontent* (New York: The Free Press, 1964).

7. This essential utilitarianism is an underlying theme of Danny Wynn Ye Kwok's *Scientism in Chinese Thought, 1900-1950* (New Haven: Yale University Press, 1965). It is also consistent with Lucian Pye's suggestive, but overstated observation that

> . . . the Chinese political culture has never recognized that any autonomous bodies of knowledge can properly lie outside the reach of the formal political ideology.

See his *The Spirit of Chinese Politics* (Cambridge: MIT Press, 1968), p. 28.

8. See Mao Tse-tung, *Selected Works of Mao Tse-tung* Vol. II (Peking: Foreign Language Press, 1965), pp. 380-381.

9. "Speech at the Inaugural Meeting of the Natural Science Research Society of the Border Region," quoted in *New China News Agency*, January 19, 1967, in U.S. Consulate General, Hong Kong, *Survey of China Mainland Press (SCMP)* 3866.

10. Chu-yuan Cheng, *Scientific and Engineering Manpower in Communist China, 1949-1963* (Washington: National Science Foundation, 1965), p. 10.

11. Ch'en Po-ta, *Speech Before the Study Group of Research Members of Academia Sinica* (Peking: Foreign Language Press, 1952).

The extensive discussion given to Ch'en's speech below, is justified by (1) the comprehensiveness of Chen's statement of the Party's views of science and (2) Ch'en's own importance in Chinese politics. Ch'en is not a natural scientist but was a vice-president of CAS since its founding. At the time of this speech, he was also deputy director of the important Propaganda Department of the Central Committee of the CCP and a full member of the Central Committee.

The most significant aspect of Ch'en's career was his close relationship with Mao. During the Yenan period, he became Mao's personal secretary and is believed to have participated in the drafting of Mao's essays and speeches. Since the founding of the People's Republic, he is believed to have been a chief editorial writer for the official *People's Daily*. See, Howard L. Boorman (ed.), *Biographical Dictionary of Republican China* Vol. I (Ai-Ch'i), (New York: Columbia University Press, 1967), pp. 221-223. Ch'en's star continued to rise after this initial period as he became editor of the Party's theoretical journal *Red Flag* when it was founded in 1958, a member of the Politburo and a leading member of the Cultural Revolution Group. He was purged in 1971.

12. An ideological reform committee was set up by Kuo Mo-jo, President of CAS, in November, but the campaign did not begin until December 29. It was to last only four months but was interrupted by the "three anti" and "five anti" campaigns. See U.S. Consulate General, Hong Kong, *Current Background* (hereafter *CB*) No. 153.

13. Ch'en, op. cit., p. 3.

14. Ibid., p. 5.

15. Ibid., pp. 7-8.

16. Ibid., p. 13.

17. Ibid., p. 14.

18. The terms "internalist" and "externalist" are from the historiography of science. The "internalist" sees scientific development as the product of *internal* logic of science itself, and rather independent of sociological or economic factors.

19. See Joseph Needham, "Science and Society in East and West," in Maurice Goldsmith and Alan Mackay (eds.), *Science and Society* (New York: Simon and Schuster, 1964), pp. 147-149. This question of the internalist vs. externalist explanations of the rise of science surely is not only relevant to China but also to all "late modernizers."

20. Ch'en Po-ta, op. cit., p. 26.

21. Ibid., p. 26.

22. "On the basis of the above-mentioned orientation and various problems explained thereunder, we should be able to map out some research plans." Ibid., p. 26.

23. Ibid., p. 16.

24. Ibid., p. 21.

25. Ibid., p. 20.

26. Ibid., p. 21.

27. "Discover the truth through practice, and again through practice verify and develop the truth. Start from perceptual knowledge and actively develop it into rational knowledge; then start from rational knowledge and actively guide revolutionary practice to change both the subjective and objective world. Practice, knowledge, again practice, and again knowledge. This form repeats itself in endless cycles, and with each cycle the content of practice and

knowledge rises to a higher level. Such is the whole of the dialectical-materialist theory of the unity of knowing and doing." *The Selected Works of Mao Tse-tung* Vol. I (Peking: Foreign Language Press, 1965), p. 308.

28. Ch'en, op. cit., p. 23.

29. Ibid., p. 28.

30. Ibid., p. 29.

31. Mao, *On Coalition Government*, quoted in Ch'en, p. 30.

32. Ch'en, op. cit., pp. 7-8.

33. See Yü Kuan-yüan, "Two Roads for Developing the Scientific Enterprise in China," *Jen-min Jih-pao*, September 4, 1957, in *SCMP* 1620.

34. For a further discussion of the Party's doctrine for science, see Richard P. Suttmeier, "Party Views of Science: The Record from the First Decade," *The China Quarterly* 44 (October-December 1970), pp. 146-168.

3 Organization-Building for Science, 1949-1957

. . . a political system that decentralizes political power, and places it as much as possible in the hands of local units, is likely to be more successful in mobilizing the society for development and getting it to the point of accepting changes required for progress.

—Philip E. Jacob

The Chinese leaders in 1949 had a rough vision of the role of science and technology in the new society. The vision was built upon a strongly *externalist* understanding of scientific development and was one that gave greater weight to the importance of social organization than to individual creativity in explaining innovation.

The Chinese leaders also had the will to channel scarce resources into scientific activities and to devise organizations they hoped would bring about a new national scientific tradition consistent with their ideology. In translating vision into program, the Chinese relied heavily on Soviet experience and advice, particularly in their opting to build a strong Academy of Sciences as a scientific center. Since the history of scientific development in the People's Republic is intricately linked with this basic strategic decision, this chapter will explore in some detail Chinese efforts at organization-building for science in the 1949-1957 period.

Heritage of Science Organizations

Organizations for developing modern science in China go back to the 1860s when courses in astronomy, physics, mathematics, chemistry and other sciences were taught at the Shanghai and Foochow arsenals and at the T'ung-wen Kuan.[1] Jesuit missionaries established a meteorological station at Hsü-Chia-hui (Zikawei) near Shanghai in 1872 and later established a seismological station.[2] Optional questions in mathematics became part of the traditional civil service examinations in 1887.[3]

More authentically modern scientific organizations did not appear until the establishment of the Chinese Republic in 1911 and then only slowly. The first was the Geological Survey of China, founded in 1916 under the Ministry of Industry. Various non-government organizations fostered science by establishing research centers, such as the biological institute at Nanking, sponsored by the

Science Society of China in 1922,[4] and the Fan Memorial Biological Institute in Peking, founded in 1928 with Boxer Indemnity funds by the China Foundation for the Promotion of Education and Culture.[5] The major scientific research organizations of the Republican period, however, were the Academia Sinica (*Chung-yang Yen-chiu Yuan*), founded in 1928, and the Peking Academy (*Pei-ching Yen-chiu Yuan*), founded in 1929.[6] These latter two organizations formed the nucleus for the new Chinese Academy of Sciences (*Chung-kuo K'o-hsueh Yuan*) established by the Communist regime in November 1949.

The Initial Period of the Chinese Academy of Sciences (CAS): 1949-1953

At the beginning of the new regime, CAS was to have two main functions. First, it was to be an institution actively engaged in scientific research primarily to serve the national goals of economic development and increasing military capability. In addition, CAS was to have an administrative function as that agency of the government in charge of all matters relating to science and to social and historical studies.[7] This latter function included such activities as (1) conducting a national manpower survey and planning for the rational distribution of scientists geographically and by sector;[8] (2) urging Chinese scientists studying or conducting research abroad to return to China; (3) coordinating the activities of CAS with those of institutions of higher education, particularly with regard to the training of new generation of scientists and to the place of employment of experienced scientists; (4) conducting an ideological reform campaign among CAS personnel;[9] and (5) controlling scientific information–the standardization of terminology, translating foreign scientific publications, and preparing bibliographies of science books.[10]

From the CCP point of view, such tasks would require that the inadequacies of the old research system–its lack of coordination, duplication of effort, poor financing, and separation from the needs of society–be overcome. Accordingly, throughout 1950 some 38 conferences were held on the establishment of a Chinese Academy of Sciences that would incorporate the facilities remaining from the two old academies. The resulting organization of CAS for research consisted of fifteen research institutes. These were subsumed under five groups–physical science (two institutes), chemical science (two institutes), earth science (two institutes), biological science (five institutes), and social science (four institutes). In addition, preparatory offices were established for three more institutes. The Tzu-chin Shan (Purple Mountain) observatory at Nanking was also subsumed under CAS.

Formal academic leadership in the new Academy rested with a president and five vice-presidents. A General Office was to handle administration.[11] Members of the new Academy were recruited in a reasonably simple procedure. From the old Academia Sinica membership, 35 scientists were chosen as electors of new members. This group then chose new members in various disciplines, selection presumably being based on achievement. This initial selection resulted in a membership of 233. In March 1950, a second election was held, which enlarged the membership to 865 persons. Some 174 of these scientists were elected *in absentia*, since they had chosen to stay abroad, at least at that time.[12] From the available group of 691, the Government Administrative Council selected the directors of research institutes.[13] The structure of CAS was further differentiated with the establishment of the Academic Review Committee (*Hsueh-shu Ping-shen Wei-yuan-hui*), a group composed of some 26 scientists of "high academic standing" who were to discuss the basic business of the Academy.[14] The initial committee later gave way to an Academy Council (*Yuan-wu hui-yi*), which more nearly reflected the General Meeting of the Soviet Academy of Sciences. The Academy Council was a large group of several hundred scientists who were to meet not more than once a year. The body that compared in size and function with the Presidium of the Soviet Academy was the Standing Committee of the Council (*Yuan-wu chang-wu wei-yuan-hui*). In addition, a "Committee of Specialists" (*Chuan-men Wei-yuan-hui*) was organized into some twenty sections, reflecting the division of the institutes, to review purely scientific, as opposed to administrative, matters. Membership in each section ranged from 3 to 18 people and included substantial numbers from universities, government departments, and industrial establishments.[15] The committee was to advise CAS on research matters and was an early attempt to establish—and keep open—channels of communication between the chief performer and planner of the nation's research (CAS) and the units that were functionally related to it (factories, universities, and so forth).

In this early period, then, the responsibility for planning and implementing a national science program rested almost exclusively with CAS. It was not only responsible for its own operations but was to play a coordinating and leading role *vis-à-vis* the university and industrial research that was anticipated. What research planning that was done was carried on through the office of the CAS president in consultation with government departments. The responsibility for distributing agreed upon research tasks rested with the vice-presidents.[16]

Although some significant scientific work was done in this period—particularly in geological and biological surveys—research was generally limited. Time not given to organization was spent on ideological reform.[17] The considerable amount of time spent to organize and ideologically orient the Academy, however, was not seen as a waste. Physicist Ch'ien San-ch'iang, probably reflecting the views of the regime, argued that the attention given to non-research activities should be viewed as an investment that would pay off in the future.[18] It seems, however, that the initial period of organization was no sooner over, than the inadequacies of the organization became apparent.

The Period of Major Changes: 1954-1957[19]

By 1954, two interrelated developments in China contributed to the major changes in the Chinese science system that began in that year. First, China had already begun the program of ambitious industrialization outlined in the First Five-Year Plan (FFYP). Chinese national leaders obviously had expected the science system, headed by CAS, to be a ready source of innovation for economic production activities. CAS was not structurally prepared, however, nor were CAS scientists temperamentally prepared for such a task. Thus, the demands of the FFYP and CAS' inability and unwillingness to meet those demands would necessitate a period of change.

Not only was the environment ripe for change, then, but agents prepared to precipitate and direct change were present. This second development, the presence of Soviet advisers who were coming to China in increasing numbers by 1954, also had a major effect on the system. Faced with apparent Chinese dissatisfaction with China's science system, the Russians responded by recommending reorganization along the lines they knew best.

From 1954 to 1957, organizationally and administratively, CAS was almost continually occupied with the following activities: the creation of an administrative Secretariat to supersede the General Office; the establishment of "academic departments" (*hsueh-pu*), modeled after those in the Soviet Academy of Sciences; the authorization of scientific awards by the State Council to be given by CAS; the enrollment of post-graduates for advanced training; and the full-scale introduction of planning in science. In addition, the number of research institutes, and the number of personnel under CAS increased steadily.[20] Financial support of CAS also was expanded.[21]

An early call for major changes was a report on the work of CAS given by President Kuo Mo-jo to the 204th Government Administrative Council (GAC) meeting in January 1954.[22] Kuo began by touching on two basic questions of Chinese scientific development—manpower supply and autonomy of CAS. He noted that from the narrow point of view of CAS scientists, things did not appear to be too bad. There was, to be sure, a shortage of trained manpower, but there were also a number of eminent scientists who could carry on research. The scientists seemed to be saying that given time and autonomy the number and quality of scientists would increase. Kuo noted that from the point of view of the needs of national construction, on the other hand, the situation, particularly in certain underdeveloped fields of science, was critical. According to Kuo, there should be greater coordination of theory and practice, and increased cooperation and unity among scientists. These deficiencies point to the need for improved leadership within CAS. Specifically, this leadership should organize the scientific elite to give guidance on academic matters, make provisions for academic activities such as conferences and discussions, and reduce the heavy administrative burdens borne by heads of institutes.[23] Kuo's report indicated that the structures provided for in the initial organization of CAS were not performing their essential functions. He therefore suggested the following measures:

1. A special organ in the State Planning Commission should be established to examine research plans in order to avoid duplication and waste of effort and equipment, "to enable the total needs of the nation to be organically and closely coordinated with the strength of the various research institutes." (This first proposal resulted in the establishment of the Scientific Planning Commission, in March 1956, to be discussed below.)
2. CAS institutes of technical sciences should coordinate their activities further with industrial units.
3. A Secretariat should be set up under the Academic Council to assist in administrative matters. Preparations should begin for establishing academic departments (*hsueh-pu*).
4. Each institute should have an academic committee (*hsueh-shu wei-yuan hui*) composed of the institute head and selected scientists. These committees should draw up institute research plans, and guide academic discussions. They should be assisted by a deputy director of administration.[24]

The Secretariat (Hsueh-shu mi-shu ch'u)

The first concrete step taken to implement Kuo's suggestions was the establishment of the Secretariat on April 8, 1954. Information about the Secretariat is scarce, but we know that its establishment was regarded as a first step towards developing a Soviet-style academy.[25] In the Soviet Academy of Sciences at that time, the Scientific Secretary had the dual function of insuring that plans were fulfilled (which included control over recruitment of personnel) and of providing services (clerical staff, motor pool, and so forth) for the Academy.[26] Similar functions of the Chinese Secretariat can be inferred from the various bureaus and offices under its jurisdiction. These were divided into those with administrative missions—the bureaus of administration, personnel, liaison, foreign relations, scientific personnel training, and planning—and those with more scientific missions—Committee on Translation and Publication, Comprehensive Expedition Committee, Science Scholarship Committee, Institute of Scientific and Technological Information, and other specialized committees. The former group tended to be led and staffed by party non-scientists, while the latter group was led by people with *bona fide* academic backgrounds.

A classic example of how important these subsecretarial units were and how they operated (or failed to operate) came to light during the spring of 1957 when scientists met to discuss "contradictions within their ranks." Chemist Yüan Han-ch'ing (Ph.D., University of Illinois) was given the responsibility of organizing and heading the Institute of Scientific and Technological Information. This task required contacts with both the Personnel Bureau and the Administration Bureau. The former controlled the allocation of personnel for the new institute, while the latter controlled material requirements—premises, equipment, and so forth. Yüan reported that he was given a continuing run-around by these two bureaus since no personnel would be forthcoming until the question of premises was settled. Likewise, no plans could be made to set aside space and buildings until there was some indication that the new institute would have personnel.[27]

The relationships between the Secretariat and the academic leadership in CAS is difficult to determine. In the Soviet Union, the center of authority within the Academy has been located somewhere between the Presidium and the office of the Chief Scientific Secretary, its precise locus being dependent upon the personalities of the President of the Academy and the Secretary. In China, until the Cultural Revolution at least, the balance of authority was tipped consistently in favor of the Secretariat. The Secretary General (*mi-shu chang*) and Deputy Secretaries-General have been known more for party affiliation than for scientific achievement. Physicist Ch'ien San-ch'iang—prominent scientist, party member, and administrator—was a notable exception.[28] Interestingly however, while Secretaries- and Deputy Secretaries-General have come and gone over the years, the academic leadership has remained remarkably stable.

The Departments

Following closely behind the establishment of the Secretariat was a decision to prepare for the establishment of a system of Soviet-style departments, announced on April 13, 1954.[29] These preparations were to be headed by four committees—one for each of the departments: (1) physics, mathematics and chemistry; (2) biology and geology; (3) technical sciences; and (4) social sciences.[30] This preparation took more than a year and it was not until June 1, 1955, that the inaugural meeting of the departments was convened.

The establishment of the departments was apparently part of the Sino-Soviet strategy of creating parallel institutions for facilitating mutual cooperation. But there were also special problems within the organization of Chinese science that the departmental system would help ameliorate. There was, for instance, an acknowledged weakness in academic leadership in the old system.

According to Kuo Mo-jo:

> . . . the main defect in the leadership work of the Academy is failure to conduct research into the needs of national construction so as to formulate plans for developing our scientific work. Our contact with government departments has been far from adequate and we are not clear about the demands of government departments on scientific work. Nor are we clear about the developments in various branches of science and the potentialities of our scientists. We have not yet brought forward a long-range plan for the development of science in China. We are still not able to define, on the basis of the demand of national construction, what scientific departments we should develop, what important subjects we should study, what research organs we should set up, what research

organs should be set up first, what research organs should be set up afterwards, where research organs should be set up, how many research personnel are required, where they should come from, and how they should be trained.[31]

Furthermore, the problem of leadership seemed to come from the failure of CCP "leadership" cadres in CAS. With a few exceptions the cadres were untrained scientifically. They did not engage in research and had little to say about academic matters.[32] China seemed to be experiencing a science policy breakdown, in which the broad directions of policy could be stipulated by high-level non-scientists in government but in which such directives were non-operational when decisions about priorities in research had to be made at lower levels. If the basic tenet of Chinese science policy was that research should serve production, would production be best served by concentrating resources in biochemical research or in research on high polymers? Lower level Party cadres were totally unequipped to make these scientific choices, and policy from the top was manifestly lacking in specificity.[33] One purpose of the department system was undoubtedly to remedy this situation.

Two hundred thirty-three of China's most accomplished scientists were chosen as board members[34] to make up a Department Committee (*hsueh-pu wei-yuan hui*). At first this committee was to meet in plenary session twice a year, but this was reduced to once a year in 1957.[35] In addition to the Department Committee, there was also a 71-member Standing Committee (*hsueh-pu ch'ang-wu wei yuan hui*), which was elected by the board members and which was to meet once a month. The composition of the Standing Committee again reflected the prevalence of achievement criteria. The composition of the Department Committee was not limited to those *employed* at CAS; it also included university professors, engineers, and researchers from production departments.[36] The inclusion of non-CAS personnel seems to point to three other problems within Chinese science that the establishment of the departments would resolve.

The first relates to the changed status of CAS following the new 1954 constitution of the Chinese People's Republic. Prior to 1954, CAS had ministerial status. Following the new constitution and the organic law of the State Council, CAS no longer was a component part of government, but "the supreme academic organization under the State Council." While CAS' responsibilities from the beginning extended beyond its own institutes and personnel, its new status made it quite clear that if necessary, organizational changes should be made to insure that non-CAS research institutes were heard (promoting "academic democracy") and that CAS "gradually became a center to unite the scientists of the entire country and to lead and push forward scientific work."[37]

Secondly, if CAS was "the supreme academic organ of the State charged with the important mission of directing and promoting scientific work throughout the country"[38] and if the main purpose of research was to aid national development, then it would be necessary to have institutionalized communications between the various research sectors and the users of research in production settings. The departments would thus be expected to contribute to this joining of theory and practice.

Although much of the discussion around the establishment of departments concerned "academic leadership" (*hsueh-shu ling-tao*), the most concrete discussions of academic leadership pertained to *planning* in science. Thus, the third, and perhaps most important, reason for establishing the departments would be to make them the instruments for instituting research planning. The departments would aid in transforming the general policy directives regarding the goals of research into concrete "scientific choices."

The inaugural meeting of departments was not only a celebration but also a working meeting at which the outlines for CAS' First Five-Year Plan were set.[39] Three months later, in September 1955, the departments were instructed to begin work on drawing up a long-range scientific development plan.[40] In addition, the director of each department was to submit quarterly and annual plans to the Department Committee or to its Standing Committee. Annual plans required the approval of the central CAS leadership.[41]

The department system, particularly in relation to planning, represented a recognition by the CCP that if science were to be planned and organized, it would be essential that a significant number of scientists be involved in the planning process.

> In order to strengthen leadership over scientific undertakings, it is necessary to study practical work, to raise vital questions as regards academic thought and methods of research to point the direction of development. For this reason, a certain number of scientists well-versed in scientific undertakings must be relied upon to do this work.[42]

However, the department system was not a simple step towards the advancement of professional interests. By bringing in non-CAS personnel, and by continuing to emphasize the pragmatic importance of research,[43] there was no possibility for CAS to insulate itself from non-professional demands.

Apart from involvement in planning, the departments were also charged with solving the day-to-day problems of research, both in CAS and in universities and industry; promoting cooperation among research organs; checking work of subordinate research organs and hearing their work reports; checking up on training of personnel and on the scientific publications of CAS; reviewing research from below to see if it had economic value; and holding scientific meetings and discussions.[44] At least once a year, the departments were to report on the work of their subordinate institutes to the Academy Council.[45] Each department had a director and one or more deputy directors. Routine matters were handled by a standing committee consisting of the director, the deputy(ies), and selected members of the department. Assisting the director were a scientific secretary and associate secretaries.[46]

Two notes should be added here about the establishment of departments. First, in spite of the existence of scientific secretaries paralleling the academic structure, the departments inevitably added to the administrative burdens of China's senior scientists. Paradoxically, in his report to the inaugural meeting, Kuo Mo-jo noted that one of the problems with CAS prior to the department system was that scientists had insufficient time for research.[47] (These burdens would later be sharply denounced by the scientists during the 1956-1957 "Hundred Flowers" period of liberalization.)

Second, the departments represented a major organizational step toward the kind of research tradition China's leaders wanted for their country. Specifically, they wanted to foster a system of organized, planned, and communal or team research and at the same time eradicate "individualism" in science that was associated with bourgeois ideology. Thus, it is not surprising that a major theme in the Chinese press at the time of the inaugural conference was that the departments would "mobilize" and "unify" the nation's scientists.

Academic Committees

The departments were a major contribution to settling the permanent organization of CAS. One further step, however, was still to be taken. This was the establishment of institute-level "academic committees" (*hsueh-shu wei-yuan-hui*). The academic committees were also called for in Kuo Mo-jo's 1954 report to GAC, but the first such committee was not set up until October 1955, when one was established at the Petroleum Research Institute.[48] By the end of 1955, only seven institutes had academic committees.[49] Judging by the slow process of establishing these units and by the following description of their composition, it is likely that scientists showed considerable opposition to them. The units were, after all, non-professional organizations imposed upon the scientists' day-to-day activities.[50]

The average number of academic committee members seems to have been 17 to 20, including the institute director and deputy director, other responsible personnel of the institute such as the scientific secretary and his assistants, representatives of Party organizations, and selected scientists. The scientists included not only the leading scientists of the institute, but also those from relevant units in universities and ministries.[51] The system of concurrent positions, which characterized organizations at the department level, was thereby established down to the institute level.

The stated purpose of the academic committees, as with the departments, was to strengthen academic leadership at the working level. More specifically, this entailed (1) studying the direction of research, examining plans for research and for cultivating new cadres, and "summing up" scientific work; (2) discussing reports of research at other research establishments whose work was relevant to the institute; (3) discussing and evaluating treatises written by researchers within the institute; (4) examining the practical usefulness of the institute's work and the possibility of its being popularized, and reporting these findings to the Academic Council; (5) making preliminary examinations of post-graduate theses (see below), promoting and grading personnel, and determining those deserving special recognition; and (6) encouraging the study of Marxism-Leninism and "advanced Soviet experience."

With the establishment of academic committees at the institute level, and specialized laboratories under the institutes the basic structure of the Academy was settled. The organization is summarized in Figure 3-1. Specialized research groups would be found below the laboratory level. For instance, below the laboratory of inorganic chemistry, under the Institute of Chemistry in Peking, which in turn would be under the Department of Physics, Chemistry and Mathematics, one could find a research group on stable isotopes.[52]

As in the Soviet Union, there was dual control over the Academy. Administrative and political control flowed from the Secretariat, scientific secretaries, and administrative committees attached to units at the various levels of CAS organization. These people may or may not have had some scientific training, although with a few exceptions, personnel in the Secretariat branch were not accomplished scientists. Similarly, deputy directors at the institute level were often low-level scientists whose chief qualification was political reliability. Many of these people, including those in the Secretariat, were Party members.

Further political control (particularly responsibility for ideological education and remolding and for mobilizing scientists for campaigns) flowed through the CCP organization, which paralleled the structure of the Academy.[53] The lowest rank in the Party organization was probably the Party cadre attached to CAS at the laboratory level. The Party apparatus extended upward through the office of the Chief Party Secretary for the Academy, to the science section in the important Propaganda Department of the Politburo, which prior to the Cultural Revolution was headed by Yü Kuang-yüan.

Planning

Along with the reorganization discussed above, another response to the desire to keep scientific development in step with economic development was the introduction of major planning in science. Vucinich has noted that in the Soviet Union, planning in science has been a major control device used to bring scientific activities under the control of the state bureaucracy.[54] Graham has described the trauma of having planning imposed on the Soviet Academy in the early 1930s, when planning in science was a very novel idea indeed.[55] By the early 1950s as opposed to the early 1930s, there was a much wider acceptance of the proposition that at least in some respects research could be rationalized,[56] although most non-Marxist scientists would still oppose state planning of research, as in fact they did in China.

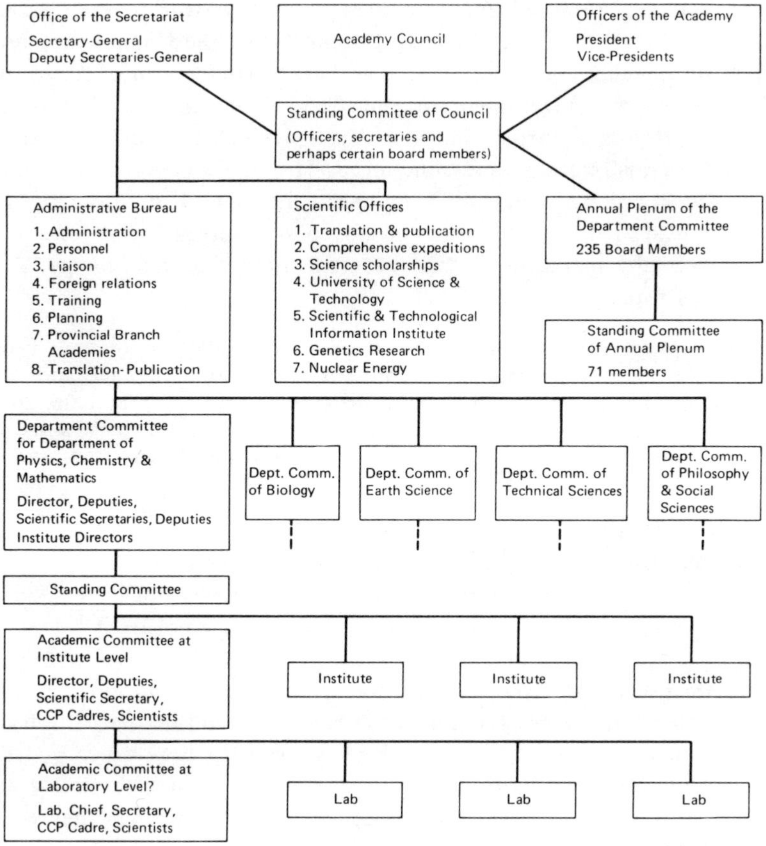

Figure 3-1. CAS Organization for Leadership and Administration. Adapted from John M.H. Lindbeck, "Organization and Development of Science," in Sidney H. Gould (ed.), *Sciences in Communist China* (Washington: American Association for the Advancement of Science, 1961), p. 35-42; and "Organization et Developpement de la Science en Republique Populaire de Chine," *Notes et Etudes Documentaires* 3255 (18 janvier, 1966), Paris La Documentation Francaise, Secretariat General du Governement, p. 13.

CAS' First Five-Year Plan

As part of the effort to accelerate scientific development following Kuo Mo-jo's report to GAC in early 1954, CAS began developing its First Five-Year Plan. This planning was apparently going on apace with the planning for departments, since on the second day of the conference to inaugurate departments (June 2, 1955), a draft of the plan was presented.[57] At the close of the eleven-day conference, it was reported that the plan was to be sent to the "Academic Affairs Committee" (i.e., the Standing Committee of the Council) for revision according to the recommendations made at the department conference. It would then be sent to the government for approval.[58] On July 24, slightly over a month later, it was reported that the Academy had announced its five-year plan.[59] Eleven major tasks were announced. The topics covered indicated that the plan did not depart radically from the work of CAS in the early 1950s, which had placed heavy emphasis on resource surveys. The one notable exception was the inclusion of "the peaceful uses of atomic energy."[60] The breakdown of topics was:

Survey-type programs (comprehensive research on establishing new iron and steel installations; oil prospecting and the study of artificial and natural petroleum; study of earthquakes and distribution of seismic belts in areas marked for industrial development; long-term planning for harnessing major rivers; study of tropical plant resources and soils of South China; demarcation of natural and economic regions of China).

Peaceful uses of atomic energy.

Social science topics (philosophic and economic problems in China's national construction during the transition period; modern and contemporary history).

Other natural science topics (antibiotics; high polymer chemistry).[61]

It apppears that the five-year plan was no sooner announced than it began losing its significance. Once the departments were established, attention shifted to the formulation of a long-term, twelve-year plan for the science system as a whole, and there are indications that many scientists ignored the five-year plan because they felt that once the twelve-year plan was formulated, they would be asked to divert their efforts towards its fulfillment.

The Twelve-Year Plan

The development of the twelve-year plan was to planning what the establishment of departments was to the structure of CAS. Indeed the two were functionally interrelated, one of the main purposes of the departments being to assist in planning.[62] Although long-range planning was discussed at the inaugural meeting, it was not until September 15, 1955, that the 39th Standing Committee of the Academy Council of CAS officially issued a directive to begin work on the plan in October.[63]

The overall objective of making science serve the needs of the state seems to be the point of departure for the plan; hence it was to be closely coordinated with economic planning (3 five-year plans). There was strong emphasis on developmental thinking in the planning. During the period covered by the plans (1956-1967), scientific and economic development would yield major changes in the society. The plan was both to make provisions for these initial changes and to take them into consideration in planning for developments in the latter part of the planning period. At CAS itself, the plan was not to be based simply on the present manpower supply, present number of institutes, or present budget. Rather, it was to include provisions for increasing the number of institutes; at the same time, in setting research goals for the 1960s, it was to take into account the increased capabilities that would come with more institutes.

The plan was also to take into account anticipated regional dispersion of industry. The intention was to overcome the regional "irrationality" of the existing system (having only a few centers of research) by planning research centers near new industrial centers. The plan was not limited to CAS in this regard but also was to include universities and industrial research units, it was to improve coordination among the three sectors.

The plan had as fundamental objectives both "balanced development" in all fields and the attainment of international standards of development after twelve years. The achievement of the latter aim required that priorities be set and hence that the former objective be redefined. Increasingly, it seemed that "balanced development" meant development in those areas where Chinese research was weak. These fields turned out to be the leading "hot" areas of world science, ones that would certainly be prerequisites for sophisticated military technology. (See the discussion of priorities later in this section.)

Rather specific instructions were given on what would be expected from scientists and administrators. For every research problem included in the plan, the following information was to be provided: (1) the significance of the problem for economic construction and/or the development of science; (2) the material conditions necessary for work; (3) the sequence of steps to be taken and the units inside and outside of CAS that would be involved or consulted; (4) estimates of time required to finish; and (5) the way in which the project would be organized.[64]

Another specified aspect of the plan concerned the training of new researchers. According to the directive, it would be necessary to determine the number (and plan for the training of those determined) of researchers to be produced within five years with the standard of "assistant research worker" (*fu yen-chiu yuan*). This number was to be broken down into those to be trained on-the-job in the Academy, those to be trained among research students, and those to be trained abroad. The plan estimated a demand in 1967 for 2 million high-level specialists with college diplomas.[65] Planning for manpower requirements was also to include the number of college graduates needed every five years, and the kinds of specialized courses the institutions of higher learning would be expected to provide. These estimates were to include a number of "auxiliary" (*fu chu*) scientific workers, or technicians, needed every five years.[66]

A final part of the plan was concerned with fiscal estimates. These particularly pertained to capital construction[67] but also to wages, apparatus, and supplies.

The planning procedure involved over 300 scientists by May 1956[68] and was to be conducted in accordance with the principle of "academic democracy"—that is, the planning process was to be open to non-CAS scientists as well as CAS personnel. The directive to commence planning contained a reasonably explicit timetable, which although not kept, nevertheless gives considerable insight into the planning operation.[69]

Beginning in October 1955, planning discussions were to be held at the institute level. At the same time, similar discussions were to go on at the department level. By December, draft plans from the institute-level discussions were to be submitted for department-level study, discussion, and synthesis into comprehensive plans for the departments. The refined drafts would then be presented to the Standing Committee of the Department Committee (*hsueh-pu ch'ang-wu wei-yuan-hui*) for review. By early January 1956, the drafts were to be submitted to the Secretariat.

The Secretariat meanwhile was to study past experiences with the planning of science from at home and abroad. By mid-January, the Secretariat, headed by a special work team which included some scientists, was to summarize and balance the drafts received from the department-level reviews and produce a long-range draft plan for the whole Academy that would then be submitted to the Standing Committee of the Council sometime during the first quarter of 1956.[70] At each stage in the refinement, the existing situation in the various fields of research in China was to be compared with advanced international levels. It would be necessary to decide what the basic scientific problems were that stood in the way of reaching advanced levels. The steps required to reach those levels could then be mapped out, including those pertaining to education as a key input into the achievement of the plan. Ultimately, the plan was to leave CAS for consideration by state and Party organs and be presented to the Eighth Party Congress in September 1956.

After going to high state and Party organs, the plan was apparently sent back down to CAS for further discussion.[71] The plan was eventually sent to the Soviet Academy for a detailed review in the fall of 1957, a review which was to be the basis for Soviet aid in fulfilling the plan.[72] It was not until late 1957 that a high-level Chinese scientific delegation left for the Soviet Union to negotiate a scientific assistance agreement based on the plan. The agreement was not settled until early 1958.

The plan itself has never been made public. It is known that it contained many projects, and that clear priorities were set.[73] The setting of these priorities effected the balanced development mentioned above.[74] The priority projects fall into twelve categories:

1. Peaceful uses of atomic energy.
2. Radio electronics (super high frequency; semi-conductors).
3. Jet propulsion.
4. Automation and remote control.

5. Petroleum and scarce mineral exploration.
6. Metallurgy (new processes and alloys).
7. Fuel technology ("heavy organic synthesis").
8. Power equipment and heavy machinery.
9. Problems relating to harnessing the Yellow and Yangtze Rivers.
10. Chemical fertilizers; mechanization of agriculture.
11. Prevention and eradication of detrimental diseases.
12. Problems of basic theory in natural science.[75]

These categories indicated major changes from the five-year plan and the work of the early 1950s. Most noticeably the number of survey-type operations was reduced to two categories (numbers 5 and 9 in the above list)[76] and there was no social science among the priorities. Except for number 12, the remaining categories pertained to applied sciences, relevant to the development of sophisticated, modern technology.

In addition to the introduction of planning, and the changes in the organization of the Academy, the 1954-1957 period included certain other developments pertaining to the motivation and socialization of scientists. Of greatest interest to us here are the establishment of a system of awards and a system of post-graduates at CAS.

Awards

As noted in Chapter 1, motivation within the scientific community is linked to what we have called the information-recognition exchange. Although it is sometimes argued that having one's work cited by others is the most meaningful recognition for scientists, prizes awarded for scientific accomplishment also have been a traditional mode of supplying recognition.

In China, the Academy did make some awards[77] before the "Provisional Regulations Governing Science Awards" were ratified in August 1955.[78] However, the Provisional Regulations spelled out the purpose of scientific awards, the procedure to be used in selections, and the nature of the prizes.

Article I of the regulations stated:

For the purpose of inspiring the position and creative talents of the scientific research workers and furthering the development of our scientific course for serving the construction of the country, science awards are to be made by the CAS.[79]

The stated purpose of the awards then was not only to recognize achievement, but also—along with the academician system (which was never established) and post-graduate system—was to be a way of raising academic standards.

Nominations for awards were to come from the working research units under the institutes at CAS or from universities or industrial labs. Scientists could also apply to their research units for awards at these low levels. Reviews and recommendations were then to be sent by the institute-level units to the departments where the basic screening would be done. The departmental choices were to be passed on to a scientific awards committee where some candidates would be recommended to the Academy Council for awards.[80] Three classes of awards were established:

First:	10,000 yuan, certificate of honor, gold medal
Second:	5,000 yuan, certificate of honor, gold-fringed silver medal
Third:	2,000 yuan, certificate of honor, silver medal

Since the highest salaries for scientists were in the 350 yuan per month range, winning even a third class award would be very lucrative indeed.

The first of the biennial awards were to be made in May 1956, but the work on the twelve-year plan ostensibly forced a delay. The first award winners were finally announced on January 1, 1957. Thirty-four papers (some had multiple authors) were chosen out of some 400 that were submitted.[81] Of the 34 awards, 3 received first-class prizes, 5 received second-class prizes, and 26 were granted third-class awards. Twenty-eight of the papers were done by individuals or teams affiliated with CAS. Sixteen papers were in physical sciences, 10 were in engineering, 3 were in biological sciences, and 3 were in mathematics.[82]

The selection process was not without its problems. One was determining whether the basic criterion for awards should be the paper's contributions to national construction or its contribution to the development of science.[83] This problem indicated the basic ambiguity of the awards. The stated purposes (above) fail to make clear whether the awards represented the freely given recognition of colleagues in support of the goals of the scientific community, or whether they were *organizational incentives* supporting the goals of the state.

The ambiguity surrounding the awards prompted a rather unusual statement in the *People's Daily* editorial of January 25, cited above.[84] The editorial stated that the areas of inquiry reflected in the awarded papers should not be viewed necessarily as fields given high priority by the state. On the contrary, a comparison of the topics awarded and the priorities set forth in the twelve-year plan reveals that there was a paucity of papers dealing with those subjects required for national construction and defense. Whether this editorial comment represented displeasure with the way the awards were handled or not, it still was a clear indication that the direction of research was not a matter to be handled autonomously by the scientific community.

Wages

Another factor affecting the motivation of scientists was material reward in the form of wages. The actions of the Chinese leadership on the question of wages and material conditions were similar to those of Soviet leaders in the early 1930s.[85] As science was brought into the work of achieving state goals, which required infringement on scientific autonomy, material conditions for scientists greatly improved. Part of this improvement included wage standards far above those of the average citizen. In China, wage policy for scientists was set by the wage reform of 1956. This reform effected a national increase of 14.5 percent, but "wages in heavy industries and key construction areas, and wages of high-grade technical workers, and high-grade scientists . . . will have a much higher increase."[86] Administrative personnel at research units would continue to be paid according to government wage standards, but wages of researchers would follow separate standards announced by CAS.[87] In addition, scientists would have such special benefits as access to government automobiles and opportunities for supplementing income through publication of articles and textbooks.[88] This wage policy seems to follow the pattern that is observable in the Soviet Union and in the West—that is, compensating scientists with generous wages and fringe benefits for the loss of autonomy entailed in working on state or industrial enterprise projects.

Post-Graduates

Crucial to both the economics and the sociology of science is the training of new research personnel. Economically, the shortage of experienced scientific manpower has been the main constraint on the development of Chinese science; sociologically, the training of new personnel involves socialization into scientific roles and hence is the basis for developing a new scientific tradition.[89]

Chinese manpower programs at this time had three objectives: (1) to increase the number of college graduates in the sciences and engineering; (2) to increase the number of technicians and support personnel; and (3) to increase the number of advanced or senior scientists, or those who were capable of leading and directing research independently. It was in support of this latter goal that the post-graduate program was initiated.[90]

The State Council at the same session (August 1955) at which the regulations on academic awards were adopted, approved the "Provisional Regulations Governing Post Graduates for the Chinese Academy of Sciences." The post-graduate system was to be a program of advanced training leading to the degree of associate doctor. The program was to begin in September 1955 with the enrollment of 50 students.

The post-graduate system was in response to the need for developing new scientists that was generally recognized by scientists and government leaders. Vice-Premier Li Fu-chun was cited as saying that the most pressing task in science was the cultivation of a new group of scientists.[91] At the same time, however, Kuo Mo-jo, in his characteristic role of mediator between government and science, affirmed that in regard to training new scientists "the idea of mass production cannot be entertained."[92] Rather, advanced scientific training could only be done through a type of apprenticeship system, in which a senior scientist would have no more than five or six apprentices working with him.[93]

In the post-graduate program, there were certain basic courses to be taken, which included dialectical materialism and Russian (or another foreign language) as well as relevant science courses. These were to be pursued relatively independently, under the guidance of a tutor, and were to be completed after two years of study. In addition, a thesis based on original research was required. The program would normally take four years, but could be lengthened or shortened to meet individual needs.

The requirements for enrollment were flexible as far as the educational origins of the candidates were concerned,[94] but high standards of achievement were required. After an initial screening of applications, the remaining applicants had to take an entrance examination, which included the fundamentals of Marxism-Leninism. Those who had done previous dissertations could have these considered as part of the examination. Those successful in the examinations would be placed in institutes, under a tutor. The departments would provide overall supervision of the process.[95]

The CAS program was to be of high quality and was to serve as a pilot venture, with universities expected to open post-graduate programs following CAS experience.[96] However, serious problems developed early. Some of those selected had difficulties transferring to CAS. The units to which they were attached, not wanting to lose the few highly trained personnel they had, tried to frustrate the moves by losing or withholding files, disrupting salaries, and engaging in other bureaucratic manoeuvres.[97] A more basic consideration, however, was the fact that the very problem that the post graduate system was supposed to solve—the shortage of manpower—worked to undermine the program. There were neither enough qualified applicants, nor enough senior researchers to make the program work. The second year's enrollment was to be over 300, but only 268 out of 1470 applicants could finally be admitted.[98] These were to be supervised by 191 tutors (51 of whom were department board members). Although there would seem to be enough senior men to go around (191 for 268 students), the assumption of tutorial responsibilities was seen as one more non-research activity required of senior scientists.

The post-graduate system was not the first or only advanced training program being conducted at CAS. In the early 1950s, the Academy brought in new scientists as research probationers. This system provided on-the-job research work and allowed for advancement within the organizational ranks of CAS (assistant researcher, associate researcher, senior researcher). The probationer program, unlike the post-graduate program, was not a course of study leading to an academic degree, however. Hence when CAS began the post-graduate program, it had created in effect a dual or two-track system of advanced training. The dual system was justified as being necessary if the twelve-year plan was to be fulfilled. However, there was a realization among some that given a shortage of tutors, the post-graduate system would definitely weaken the training of research probationers. Since the research probationers would be making a more direct contribution to the needs of national construction, the dual system was regarded by some as inappropriate.[99]

The internal development of the Academy described above represented the more significant organizational changes within China's science system—changes that were to establish CAS in a position of *academic* leadership within the system. However, decisions to expand research activities outside the Academy in the production sectors and in higher education as well as the dramatic increase in expenditures for science, which these decisions entailed, resulted in exposing the inherent weakness of CAS as the *administrative* center for China's science. China's response was to create a high-level Science Planning Commission to assume national administrative leadership.

The Science Planning Commission (SPC)

A proposal for the establishment of a state organ for planning scientific development was first made in Kuo Mo-jo's speech to the Government Administrative Council in January 1954. The organ Kuo had in mind was a high level group that would be responsible for coordination and planning among all research sectors and also between these and production units. In March 1956, a "Science Planning Committee" (*K'o-hsueh kui-hua wei-yuan-hui*) (hereafter SPC) was established directly under the State Council on a temporary basis for this purpose. The SPC was initially chaired by Vice-Premier Ch'en Yi. The four Vice-Chairmen were Kuo Mo-jo; Li Ssu-kuang, Vice-President of CAS and Minister of Geology; Po I-po, Chairman of the State Económic Commission; and Li Fu-chun, Chairman of the State Planning Commission. In addition there was a membership of some 35 scientists and non-scientists.

In December 1956, at its third enlarged meeting, the SPC was made a permanent organ of the State and its tasks were more clearly spelled out. This meeting, held while the twelve-year plan was still being formulated, was chaired by the future permanent chairman of SPC, Nieh Jung-chen.[100] At the meeting, the main tasks of the SPC were defined. These were: (1) the supervision and implementation of scientific planning, particularly, insuring that key tasks are carried out by the various sectors; (2) the summing up and balancing of the annual and long-term scientific plans and the coordination of these with the state economic plans; (3) the solution of the important problem of coordinating the different sectors of the nation's research effort; (4) the solution of problems relating to the material conditions for research—books, materials, instruments and capital construction; (5) the provision of central leadership for international scientific contacts and cooperation.[101]

By the fourth enlarged meeting of SPC, on June 13, 1957, major readjustments in the science system of China had been decided upon. First, the non-CAS sectors of the science system were to receive a greater share of the responsibility for conducting research. Universities and colleges, industrial, agricultural and medical organizations were encouraged to extend their research work and become centers of research in one or more fields.[102] This decentralization was to come at the expense of CAS, whose power and influence in relation to the total system of research would be reduced. Efforts would be made to move research activities closer to production.[103] At the same time, SPC would make a concerted effort to improve the material conditions for research. At this fourth enlarged meeting, for instance, decisions were taken to greatly expand the national library system by building new national facilities in Peking and Shanghai and regional facilities in other cities, and by setting up a national catalogue. In addition, laboratory equipment would be improved and other scientific information services would be developed.[104]

SPC itself was changed significantly. It was stated that the work of the Commission had changed greatly since its establishment as a temporary organ to participate in formulating the twelve-year plan the previous year. It had become a permanent body coordinating different sectors, formulating national science policy and administering the programs of that policy. Thus, at the June 1957 meeting, SPC made far-reaching national decisions in four areas of concern to Chinese science. First, it dealt with the national organization of research and the roles the various sectors were expected to play. As noted above, this resulted in a strengthening of the non-CAS sectors and was part of a national decentralizing of research. Second, it settled the 1957 national science plan. Third, it made arrangements for improving conditions in the areas of scientific publications, the circulation of data, and laboratory equipment. Finally, the SPC made decisions about what the proper focus of, and limits to, activity in philosophy and the social sciences should be.[105]

To operate effectively in meeting its new responsibilities, the SPC changed its leadership and membership drastically. By the end of 1957, changes in the leadership structure had come to reflect the changed circumstances in the Chinese science system. The new and permanent chairman of SPC was Nieh Jung-chen. He was assisted by four vice-chairmen, a secretary-general and seven deputy secretaries-general.[106] These were overwhelmingly CCP and government personalities with no well-known scientist involved. Kuo Mo-jo remained as Deputy Chairman, but the only other representative from CAS was Wu Heng, who was a Deputy Secretary-General of both CAS and SPC. Another important figure in SPC was Huang Ching who served both as Deputy Chairman of SPC and as Chairman of the State Technological Commission (*Kuo-chia chi-shu wei-yuan-hui*), which was established in May of 1956.

The SPC continued to operate until November 1958, when it merged with the State Technological Commission to form the State Scientific and Technological

Commission (SSTC). Nieh Jung-chen became chairman of the new Commission. Three of the five vice-chairmen came from the State Technological Commission and two came from the SPC. One was Wu Heng, who became the only person serving both on the SSTC and in a leadership position at CAS.

The SSTC continued with the work of both its original constituent commissions and in addition took ultimate responsibility for the activities of the Science and Technology Association (STA), which was formed through the merger of the All-China Association for the Dissemination of Scientific and Technological Knowledge and the All-China Federation of Scientific Societies in September 1958.[107] This new organization, whose two component parts dealt with the popularization of science and technology and with learned societies, was to become more important as popularization was greatly encouraged during the "Great Leap" years and as learned scientific societies became important functional groups beginning with the summer of 1957. Hence the SSTC, by taking responsibility for STA as well as assuming many of the original responsibilities of SPC and by possessing strong political authority, was by the end of the 1950s the most important and powerful non-research body in the Chinese science system. In addition the evolution from the 1957 version of the SPC to the SSTC represented a significant structural departure from the original Soviet model of organization.

To conclude this chapter, it will be useful to explore the structural characteristic of China's science system by examining how CAS was linked to its environment at the end of this organization-building period.

CAS-University Linkages[108]

The relationships between CAS and other institutions in Chinese society have been very much conditioned by China's perennial shortage of trained manpower. Hence there has probably been a good deal of informal inter-institutional contact simply through the persons of individual scientists who hold positions in more than one institution. In the very early years for instance, it was recognized that CAS, because of its small staff, would have to share staff with universities. CAS could add to its staff by making its facilities available to university scientists for research. Likewise, CAS scientists would be available to lecture in universities. China's scientific manpower constraints thus were responsible for what appears to be an abiding characteristic of her emerging research tradition. This characteristic will be referred to as the system of "concurrent positions," a system in which scientists play functionally related roles in more than one organization. This system was particularly encouraged when CAS and the university were geographically close.[109] In addition, CAS was to help the universities by transferring to them materials, funds, and instruments. The universities in the early years, were to act as watchdogs over the activities of the various CAS institutes by having their representatives sit on the Committee of Specialists.[110] The concurrent position system was formalized with the creation of department committees and institute-level academic committees, which included representatives from universities.

Another factor that shaped CAS-university relations was the CCP's rather uncertain attitude toward research in universities. Initially, universities were places for instruction exclusively, hence the only legitimate research was in pedagogy. This began to change by 1953-1954 as a result of discussions about the role of comprehensive universities. Experienced professors began to do research either in conjunction with CAS or individually with the assistance of junior colleagues.[111]

By the late 1950s–particularly after the discussions of spring 1957–and the beginning of the "Great Leap Forward," a much different view of university research developed. As noted above, as a result of the SPC's actions at its fourth enlarged meeting in June 1957, university research was to be strengthened. This new view reflected the Party's desire to establish a close relationship among teaching and studying, research, and productive labor. Research, for the first time, was recognized as a contribution to better teaching, but both teaching and research were to serve production. It was thought that experiencing productive labor would add new dimensions to teaching and research.

The new view of university research can perhaps best be seen in its implementation at China's most prestigious polytechnical institute, Tsinghua University. By the end of 1957, Tsinghua had begun a new approach to education and hence to the socialization of new scientists and engineers. Teaching was conducted through research seminars. The topics for the seminars–which involved team rather than individual research–came from the state and were part of the national research plans. The topics would be oriented toward the needs of national construction, one of which, for example, was a series of experiments on filtration devices done for the Paotow Steel Works.[112]

The standard procedure for the interaction of CAS and universities was for agreements to be reached between CAS and the Ministry of Higher Education. One such agreement was announced in early 1956. CAS and universities were to share research personnel and were to reduce the administrative burden of scientists to facilitate this cooperation. In addition, university people would be available for CAS projects if needed and CAS personnel would be available for teaching.[113] By far the most important cooperative agreement between CAS and the Minister of Higher Education was on the establishment of the University of Science and Technology in November 1958.

This university was a joint venture between CAS and the Ministry of Higher Education. Its president was Kuo Mo-jo, and its vice-presidents were mathematician Hua Lo-keng and automation specialists Wu Ju-yang. Most of the teaching faculty were the best scientists at CAS.[114] The university was established in response to the need for high-level training in those fields that were given high priority in the twelve-year plan. Hence the thirteen departments of the university included: "nuclear physics and engineering, technical physics, applied geophysics, chemical physics, radio electronics, radioactive and radiative chemistry, thermal dynamic engineering, high polymer chemistry and physics, applied mathematics and computing techniques, dynamics, geochemistry and rare elements, biophysics and automation."[115]

Linkages with Professional Societies

In the early 1950s, China's few professional or learned scientific societies seemed to have little or no significance for the main stream of Chinese scientific progress. Shortly after the new regime was established, these societies became part of the All-China Federation of Scientific Societies, which appeared to be just another of the mass organizations used by the CCP to politicize various identifiable social groups (labor unions, women, youth, and so forth).

By 1956-57, however, the role of the professional societies seems to have changed significantly. A typical activity of the societies during this period was to sponsor conferences that became effective forums for scientists of the same discipline—but from different research sectors—to interact. When significant disciplines had no society, special conferences were held. A geophysics conference, for instance, was held in February 1957, which afforded an opportunity for scientists from CAS, the Ministry of Geology, the Ministry of Petroleum Industry and other units to "share experiences." As a result of this conference, the Chinese Society of Geophysics was formed.[116]

Professional societies also were active in publication work, often having their own journals. In addition, after the summer of 1957, they seem to have become channels of communication between the scientists and the political authorities.

The CAS and Its Environment—A Case Study

In studying Chinese scientific development, it must be recognized that statements about one discipline or area of research cannot be automatically generalized to others. The following case, therefore, is more an example of how CAS *could* interact with various components in its environment than an example *typical* of all CAS research. The example is from physics and deals with research on semi-conductors, which was a priority area in the twelve-year plan. In February 1956, the Chinese Physical Society sponsored a national conference on semi-conductors, which was presumably attended by representatives of all research sectors. This was a new area of research in China at that time, and the conference was to arouse interest in the topic and make arrangements for research in the field. One step that was taken following the conference was to establish a course on semi-conductors, run jointly by CAS and the Ministry of Higher Education, whose purpose was "to systematically introduce the fundamental knowledge of semi-conductors and transistors to concerned personnel."[117]

At this time the work on the twelve-year plan was beginning, and semi-conductors were classified as a high-priority area. With CAS and the SPC taking the lead, interest was further generated in industry and higher education. In response to the plan, Peking University in cooperation with five other schools established specialized training courses in semi-conductors.[118] With the completion of the plan, Soviet assistance was forthcoming. Soviet advisers came to the CPR, Chinese students went to the USSR and literature and sample transistors were provided. Because of the high priority of the topic it was not long before Chinese industry began to turn out products involving semi-conductor technology.[119] Also newspapers carried news of advances in the state of the art, popularizing and generating mass interest in semi-conductors.

This case, which is one of many such cases reported in the Chinese press, highlights one of the main themes of CAS-environment interaction. The operation of the Chinese science system was supposed to, and at times did break through sectoral boundaries and concentrate on tasks. Inter-sectoral contacts directed towards the completion of research missions was an objective of the political authorities throughout the 1950s. Prior to the changes in the SPC in late 1956 and in the spring of 1957, inter-sectoral contacts depended largely on the concurrent position system. The concurrent position system, however, was more a device for research planning than an instrument for breaking down inter-sectoral barriers in the interest of task-oriented work. Instead, as we shall see in the next chapter, "bureaucratism" was frustrating cooperation between sectors, as the institutions of each sector relied on self-imposed barriers to protect their interests. The SPC became the administrative organ of the state that was charged with correcting the situation in the interest of making research more directly relevant to national objectives.

Notes

1. Teng Ssu-yu and John K. Fairbank (eds.), *China's Response to the West: A Documentary Survey, 1839-1923* (New York: Atheneum, 1966), pp. 73-76. See also Cyrus H. Peake, "Some Aspects of the Introduction of Modern Science into China," *Isis* XXII (1934-1935), p. 183.

2. Peake, op. cit., p. 186.

3. Ibid., p. 183.

4. Coching Chu (Chu K'o-chen), *Culture and Education in New China* (Peking: Foreign Language Press, 1950), in U.S. Consulate General, Hong Kong, *Current Background (CB)* 153.

5. For the most accessible, comprehensive treatment of the organization of science before 1949, see Yuan-li Wu and Robert B. Sheeks. *The Organization and Support of Scientific Research and Development in Mainland China* (New York: Praeger Publishers, 1970), Chapter II.

6. An interesting picture of the organization of research in Republican China can be found in Joseph and Dorothy Needham (eds.), *Science Outpost: Papers of the Sino-British Science Cooperative Office* (London: The Pilot Press, 1948).

7. T'ao Meng-ho, "China's New Academia Sinica," *China Monthly Review* (November 1951), in *CB* 153.

8. The various research "sectors" in China include CAS, institutions of higher education, and government research conducted in the ministries or in the specialized academies, such as the Chinese Academies of Agricultural Science and Medical Science.

9. Kuo Mo-jo, "General Report on 1950 Work and Main Points of 1951 Program of the China Academy of Sciences" (approved by the 70th Meeting of the Government Administrative Council, February 2, 1951), *Hsin-hua Yueh-pao* (May 1951), in *CB* 153.

10. Coching Chu, "Scientific Research and University Education," *Ta Kung Pao* (Shanghai, October 29, 1950), in *CB* 153.

11. These top officers of the Academy were:

President

Kuo Mo-jo, recipient of M.D. degree from Kyushu Imperial University; historian and literary figure; not a member of the Communist Party at this time, but became a member in 1958.

Vice-Presidents

1. Ch'en Po-ta, Party polemicist, long-time secretary of Mao, member of the Central Committee of the CCP.
2. Li Ssu-Kuang, British-educated geologist; member of former Academia Sinica; joined CCP in 1958.
3. T'ao Meng-ho, sociologist; member of former Academia Sinica.
4. Chu K'o-chen, American-educated geographer-geologist; member of former Academia Sinica; joined CCP in 1962.
5. Wu Yu-hsun, American-educated theoretical physicist.

Head of General Office

Yen Ch'i-tz'u, French educated physicist; member of Central Committee of Chiu San Society.

This formal leadership lasted, with some minor changes, until the beginning of the Cultural Revolution. T'ao Meng-ho died in 1960. Chang Ching-fu (Party figure) became a Vice-President in 1956 and in 1958, concurrently Secretary, CCP Committee of CAS. Pei Li-sheng (another Party figure and Secretary-General of CAS, 1956-1960) became a Vice-President in 1960. Changes as a result of the Cultural Revolution will be discussed in the next chapter.

12. Ch'en Feng-chi, "Scientific and Technical Education in China," *Ajiya Kenkyu (Asia Research)* 387 (Tokyo, 10 September 1963), in the United States Department of Commerce (Washington, D.C.: *Joint Publications Research Service [JPRS])*, pp. 22 and 237.

13. Kuo Mo-jo, op. cit.

14. T'ao Meng-ho, op. cit.

15. T'ao Meng-ho, op. cit.

16. Kuo Mo-jo, op. cit.

17. Ideological reform was an important part of the lives of scientists in the early 1950s and has received serious scholarly attention elsewhere. See especially Theodore H.E. Chen, *Thought Reform of the Chinese Intellectuals* (Hong Kong: Hong Kong University Press, 1950).

18. Ch'ien San-ch'iang, "The Development of Natural Science in New China," *People's China* (September 16, 1953), in U.S. Consulate General, Hong Kong, *Survey of the China Mainland Press (SCMP)*, p. 673.

19. Perhaps realizing that organizing CAS took longer than expected, an authoritative Party spokesman, writing in 1959, defined the period of organization and orientation as from 1949 to 1956. Tu Jun-sheng, "Great Progress Made in the Natural Sciences in China During the Last Decade," *Scientia Sinica* VIII 2 (November 1959), p. 1197.

20. According to Kuo Mo-jo, in 1952 there were 31 institutes under CAS. This number had risen to 68 in 1957. In 1952, there were 1,290 research personnel (researchers, research assistants) in CAS. In 1957 the number of research personnel (including postgraduates but excluding technical support personnel) was 5,500. *New China News Agency (NCNA)*, March 17, 1958, in *SCMP* 1737.

21. During this period, the absolute level of financial support of CAS showed a steady rise. The CAS budget as a percent of the national science budget, however, declined sharply.

22. Kuo Mo-jo, "Report Concerning the Basic Conditions of the Chinese Academy of Sciences and its Future Work" (Given at the 204th meeting of Government Administrative Council, January 28, 1954), *Jen-min Jih-pao (JMJP)*, March 26, 1954, in *CB* 359.

23. Ibid.

24. Ibid.

25. Li Ssu-kuang is reported as saying that the Secretariat and the departments are concrete steps taken from the study of Soviet advanced experience. "Chinese Academy of Sciences Sets up Secretariat to Strengthen Leadership of Scientific Research Work," *NCNA*, April 14, 1954, in *SCMP* 792.

26. Alexander Vucinich, *The Soviet Academy of Sciences* (Stanford, California: Stanford University Press, The Hoover Institute Studies. Series E: Institutions, November 3, 1956), p. 22.

27. "Academy of Sciences Convenes Forum for Peking Scientists to Discuss Contradictions Within Their Ranks," *NCNA*, May 3, 1957, in *SCMP* 1541.

28. Ch'ien San-ch'iang was apparently the Academy's first Secretary-General in 1954. Ch'ien had a distinguished career as a student of Irene Joliot-Curie in France before returning to China in the late 1940s. In addition to his important role at CAS, he also taught at Tsinghua University and had been one of two leading scientists in the nuclear weapons development program. He was severely criticized during the Cultural Revolution.

Pei Li-sheng, was Secretary-General from 1956 until 1960 when he became a Vice-President of CAS. Pei is a graduate of Moscow University and was active in Party affairs going back to the Yenan period.

As of 1959, the Deputy Secretaries were Ch'in Li-sheng, Wu Heng (a geologist who was apparently one of the original Deputy Secretaries in 1954), Hsieh Hsin-ho, Yü Wen, and Tu Jung-sheng. Tu, formerly the Secretary General of the CCP Central-South Bureau, with some background in agricultural affairs and a Deputy Secretary of the Scientific Planning Commission (1956-1957) became the Secretary-General in 1961 and held the post until the Cultural Revolution.

This information comes from *Chuka Jinmin Kyowakoku Soshiki Betsu Jinmeihyo (Tables of Personal and Government Organizations of the Chinese People's Republic)*, (Tokyo: Naikaku Kambo Naikaku Chosashitsu, 1957, 1959, 1961); *Who's Who in Communist China* (Hong Kong: Union Research Institute, 1966); Cheng Chu-yüan, *Scientific and Engineering Manpower in Communist China*, 1949-1963 (Washington: U.S. National Science Foundation, 1965).

29. "Chinese Academy of Sciences to Set Up Four Departments to Guide Scientific Research," *China News Service*, April 13, 1954, in *SCMP* 788.

30. The second of these was split into separate departments of biological sciences and earth sciences in 1957. It is quite possible that this shift reflected a basic change in the substance of CAS research in these fields. In the early years, much of CAS' work generally, and particularly in earth and biological sciences, was in field surveys and expeditions. As research became more diverse and as the number of personnel increased, the old and undifferentiated administrative structure was rendered obsolete.

31. Kuo Mo-jo, "Report Delivered at Inaugural Meeting of Departments of Chinese Academy of Sciences." *JMJP*, June 2, 1955, in *CB* 359.

32. Ibid.

33. On the concept of "scientific choice," see Alvin M. Weinberg, "Criteria for Scientific Choice," *Minerva* I (Winter 1963), pp. 159-171.

34. The "board members" have been compared to "academicians" in the Soviet Union. This comparison cannot be made unambiguously, however. At the time of the establishment of the departments, there were explicit statements in the Chinese press that one of the responsibilities of the Departments was to prepare for a "system of academicians." This system was never instituted however. The board members *are* comparable to Soviet academicians in the sense that both are the scientific elite of their countries. There is no indication that

board membership was intended to be an *honorary* position, however. Rather they were to be working positions, similar in function to the officers of the Soviet departments, many of whom may be full academicians. See Cheng Chu-yüan, op. cit., p. 322.

About 60 percent of the board members in the natural sciences had earned doctorates in foreign countries. Cheng, op. cit., Appendix IV.

35. During the "Great Leap Forward," this regulation was ignored. It appears that there were no meetings between 1957 and 1960.

36. An unrefined survey of the employment patterns of 159 CAS board members (excluding those from the Department of Philosophy and Social Science) showed that only 50 were *primarily* employed at a CAS unit, while 90 were *primarily* employed at non-CAS installations. Nineteen held regular positions at both CAS and non-CAS units. Of 40 standing committee members, 15 were at CAS, 16 were not primarily employed at CAS and 9 were employed at both CAS and non-CAS facilities. The raw data for this survey are from Cheng Chu-yüan, op. cit., Appendix IV, pp. 369-572.

37. "Congratulations to the Chinese Academy of Sciences on the Inauguration of Departments," *K'o-hsueh T'ung-pao (KHTP)* (editorial), 6 (1955), in *SCMP* 1092.

38. "Important Steps to Strengthen Leadership Over Scientific Work," *JMJP* (editorial), June 3, 1955, in *SCMP* 1064.

39. "This inaugural meeting shall be a good beginning for scientists to apply their special knowledge and take an active part in formulating plans for developing the sciences." Ibid.

40. "Directive on the Drawing Up of a 15-Year Long-Range Plan of the Chinese Academy of Sciences," *KHTP* 11 (1955), in U.S. Consulate General, Hong Kong, *Extracts from China Mainland Magazines (ECMM)*, p. 20.

41. "Regulations on the Organization of the Departments of the Chinese Academy of Sciences," in Cheng Chu-yüan, op. cit., p. 304.

42. "The Beginnings of a New Stage of Our Scientific Undertakings," *Kuang-ming Jih-pao (KMJP)* (editorial), June 3, 1955, in *SCMP* 1068.

43. "Scientific Research to Help Industry, says Yen Ch'i-ts'u, Head of Department of Technical Sciences of Academy," *NCNA* June 9, 1955, in *SCMP* 1068.

44. "Congratulations to the Chinese Academy of Sciences on the Inauguration of Departments." See also "Regulations on the Organization of the Departments of the Chinese Academy of Sciences," in Cheng Chu-yüan, op. cit., pp. 302-303.

45. "Regulations . . . ," in Cheng Chu-yüan, op. cit. p. 304.

46. Ibid., p. 303.

47. "Report of Kuo Mo-jo at Inaugural Meeting of Departments of Chinese Academy of Sciences," *NCNA*, June 2, 1955, in *SCMP* 1066.

48. "Academic Committee Set Up By Petroleum Research Institute of the Chinese Academy of Sciences," *JMJP*, October 20, 1955, in *SCMP* 1162.

49. "Seven Research Institutes of the Chinese Academy of Sciences Set Up Academic Committees," *NCNA*, December 26, 1955, in *SCMP* 1201.

50. Klochko, in his capacity as a Soviet advisor to the CAS Chemistry Institute in Peking, noted in 1958 that "The Institute has no learned council of leading scientists, no curriculum committee, no cooperative council that could discuss the programs, study reports, and evaluate the findings and opinions of the individual research workers, as was done in our Moscow Institute." Mikhail A. Klochko, *Soviet Scientist in Red China* (trans. Andrew MacAndrew), (New York: Frederick A. Praeger, 1964), p. 26.

51. Ibid. The information about these committees is sketchy, but they seemed to follow the Soviet system, particularly in having both a scientist and a representative of the Party (primary Party organization in the USSR) involved. What may be a significant difference from the Soviet system is the presence of non-Academy scientists on these committees.

52. See N.P. Luzhnaya, "Scientific Institutes of the People's Republic of China," *Zhurnal Neorganicheskoy Khimii (Journal of Inorganic Chemistry)* Vol. III, No. 2 (1958), in *JPRS* 2568. See also Lindbeck, op. cit.

53. Cf. Vucinich, op. cit., p. 29.

54. Ibid., pp. 81-89.

55. Loren B. Graham, *The Soviet Academy of Sciences and the Communist Party, 1927-1932* (Princeton: Princeton University Press, 1967).

56. "At the outset it is well to distinguish between planning for research and planning of research. It is well accepted that consistently productive research must be planned for in the sense that competent men must be assembled, facilities provided, and equipment installed for their use." J.R. Steelman, *Science and Public Policy* (Washington, 1947). 28. Cited in Graham. op. cit. p. 64.

57. "Chinese Academy of Sciences Tables Draft Five-Year Plan," *NCNA*, June 3, 1955, in *SCMP* 1062.

58. "General Resolution Adopted at Meeting of Chinese Academy of Sciences, 15-Year Plan Called For," *NCNA*, June 11, 1955, in *SCMP* 1068.

59. "Academy of Sciences Announces Five-Year Plan," *NCNA*, July 24, 1955, in *SCMP* 1068.

60. Ibid.

61. Ibid.

62. Officially, the necessary or "objective" conditions for drawing up the plan existed:

1. The First Five-Year Plan for national construction was known and had been in effect.
2. A five-year plan at CAS had been worked out.
3. The nation's best scientists had been brought into academic leadership through the organization of departments.
4. A delegation from the Soviet Academy and the Soviet advisers in CAS had put forward important suggestions.

See, "Directive on the Drawing up of a 15-Year Long-Range Plan of the Chinese Academy of Sciences," *KHTP* 11 (1955), in *ECMM* 20.

63. Ibid.

64. Ibid.

65. Chu-yüan Cheng. "Scientific and Engineering Manpower in Communist China," in Joint Economic Committee of the U.S. Congress, *An Economic Profile of Mainland China* (New York: Frederick A. Praeger, Publishing, 1967), p. 522.

66. The anticipated demand in 1967 was 4 million. Ibid., p. 522.

67. "Directives on the Drawing Up of a 15-Year Long-Range Plan of the Chinese Academy of Sciences."

68. "Chou En-lai Gives Reception for Scientists," *NCNA*, May 26, 1956, in *SCMP* 1299.

69. "Directive on the Drawing Up of a 15-Year Long-Range Plan of the Chinese Academy of Sciences."

70. Ibid.

71. A news story from Shanghai, dated March 9, 1957, said that discussion of changes and revisions in the plan had been going on "during the past month." "Shanghai Scientists Discuss 12 Year Plan," *Wen Hui Pao*, March 9, 1957, in *SCMP* 1500.

72. Mikhail Klochko, who took part in these review sessions in the field of chemistry, notes that they were conducted in a haphazard fashion. See Klochko, op. cit., p. 7.

73. The Soviet Union promised help on 100 of 582 projects. Wu and Sheeks, op. cit., p. 63.

74. Kuo Mo-jo stated that top priority should be given to "those branches of science which are urgently needed in national construction but which are deficient or weak under the present conditions." Kuo went on to say that certain high priority works may be submitted for approval before overall planning is completed. This was in fact done in such fields as automation, computer technology and semiconductors. See Kuo, "Development of Scientific Research in China" (report given to 3rd Session of 1st National People's Congress, June 18, 1956), *NCNA*, June 18, 1956, in *CB* 400; and "New Scientific Institutes to Be Set Up," *NCNA*, November 1, 1956, in *SCMP* 1405.

75. *NCNA*, December 30, 1956, in *SCMP* 1442.

76. No attempt is made here to show the priorities among these 12 categories. The great costs associated with nuclear weapons development, however, would indicate that those categories related to this program (certainly 1 through 6 and 12) would rank high.

77. Individual and institutional awards at the Institutes of Mechanical Engineering, of Electrical Engineering, and of Applied Chemistry were reported in June 1955. "Chinese Academy of Sciences Grants Awards to 19 Research Units and 23 Individuals," *NCNA*, June 4, 1955, in *SCMP* 1062.

78. Ratification was by the 17th plenary session of the State Council, August 5, 1955. The regulations can be found in English in Cheng Chu-yüan, op. cit., pp. 317-319.

79. Ibid., p. 317.

80. "Regulations on Awards to Scientists," *NCNA*, September 1, 1955, in *SCMP* 1122.

81. "First Science Awards of China," *JMJP* (editorial), January 25, 1957, in *CB* 437.

82. The authors, papers, and affiliation of authors are given in "China Grants Science Awards for the First Time," *NCNA*, January 24, 1957, in *CB* 437.

83. "Explanation of the Procedure Governing the Adjustment of the 1956 Science Awards [Natural Science Section] of Chinese Academy of Sciences," *JMJP*, January 25, 1957, in *CB* 437.

84. See note 81.

85. Cf. Graham, op. cit., p. 146.

86. "State Council Decision of 16 June 1956 Regarding Wage Reform," *Kung-jen Jih-pao*, 5 July 1956, in *JPRS* 585 N.Y. 2.

87. *Kung-jen Jih-pao*, 22 August 1956, in *JPRS* 515 N.Y. 24.

88. Cheng Chu-yüan, op. cit., pp. 151-152.

89. This point was quite obvious to the Maoist during the Cultural Revolution. Both the system of CAS awards and the post-graduate system were attacked in 1967 and were attributed to the scheming of Lu Ting-yi in the mid-1950s. Lu, in turn, was charged with acting on the instructions of "China's Khrushchev," Liu Shao-ch'i. These "capitalist roaders" attempted to use both the award and the post graduate systems in ". . . an attempt to use fame and gain as bait, and cultivate the younger generation as bourgeois successors to meet his (Liu's) needs." Shih Yen-hung, "Down With The Chief Backer of the Revisionist Educational Line," *JMJP*, July 18, 1967, in *CB* 836.

90. Cheng, "Scientific and Engineering Manpower in Communist China."

91. Kuo Mo-jo, Speech to National People's Congress, July 21, 1955, *NCNA* July 21, 1955, in *CB* 351. Note that this was about one month before the regulations were adopted by the State Council.

92. In the early 1950s a modified mass production of personnel for geological prospecting was attempted using "short-course" (two years) methods. Certain geologists were later to denounce bitterly the work done in this field since 1949 saying that Chinese geology would do well to start all over again.

93. Kuo Mo-jo, op. cit.

94. Three categories of applicants were anticipated: (1) college graduates with at least 2 years of "practical" work in research, teaching, industry, etc.; (2) exceptional college graduates with no experience; (3) those who were not regular college graduates but whose capabilities could be certified as being up to the level of a college graduate.

95. This discussion is based on the regulations that appear in Cheng Chu-

yüan, op. cit., pp. 305-311; and in *NCNA*, August 31, 1955, in *SCMP* 1129.

96. "Positively Cultivate the New Source of Strength in Scientific Research Work," *JMJP* (editorial), September 6, 1955, in *SCMP* 1129.

97. "Second Group of Postgraduate Students Enrolled by CAS," *NCNA*, April 26, 1957, in *SCMP* 1542.

98. Ibid.

99. Ibid.

100. Nieh had some early experience with science and engineering in France and Belgium in 1919-20 but spent most of his subsequent life as an important figure in the CCP and in Chinese military affairs.

101. "State Council's Scientific Planning Commission Holds Third Enlarged Meeting," *NCNA*, December 29, 1956, in *SCMP* 1453.

102. This was an authoritative announcement made by Nieh Jung-chen. "Chinese Research to be Decentralized," *NCNA*, June 13, 1957, in *SCMP* 1554.

103. Ibid.

104. "More Libraries to Help Scientists," *NCNA*, June 15, 1957, in *SCMP* 1553.

105. "Conclusion of Enlarged Meeting of the Scientific Planning Commission of the State Council," *NCNA*, June 15, 1957, in *SCMP* 1564.

106. See *Chuka Jinmin Kyowakoku Soshiki Betsu Jinmeikyo* (1957), p. 119.

107. Cheng Chu-yüan, op. cit., p. 16. The function of this organization will receive greater attention in Chapter V.

108. "University" is used here for the more cumbersome "institutions of higher education." The latter actually includes "comprehensive" (multi-disciplinary) universities, polytechnical institutes, and specialized engineering, medical and agricultural colleges.

109. Chu K'o-chen, "Scientific Research and University Education," *Ta Kung Pao* (Shanghai), October 29, 1950, in *CB* 153.

110. Ibid.

111. "Professors of Peking University Start Scientific Research Work," *Kuang-ming Jih-pao*, May 8, 1954, in *SCMP* 821.

112. Liu Hsien-chou, "A Few Questions in Scientific Research Work," *JMJP* December 19, 1957, in *SCMP* 1703.

113. "High Institutions and Scientific Research Organs to Further Cooperate in Scientific Research Work," *NCNA*, January 15, 1956, in *SCMP* 1217.

114. Leo A. Orleans, *Professional Manpower and Education in Communist China* (Washington: National Science Foundation, 1961), p. 60.

115. Ibid.

116. "Geophysics Conference Closes," *NCNA*, February 9, 1957, in *SCMP* 1469.

117. Chou Pei-yuan, "China's Progress in Physics," *K'o hsueh Hsin-wen* 30 (28 September 1959), in *JPRS* 3116.

118. Ibid.

119. Ibid.

4 Alternative Models for the Organization and Administration of Chinese Science Programs

. . . contemporary societies are called upon to challenge certain forms of organization that can no longer render the services they require, because in these societies, change and uncertainty have become the constant companions of prosperity. . . .

Such periodic "shake-ups" probably help the overall system to adapt to new social goals and thus to be more responsive to the changing needs of society.

—Organization for Economic Cooperation and Development

The end result of China's organization-building efforts during the 1949-1957 period was a differentiated system of science-related institutions. Although the influence of the Academy of Sciences was somewhat diminished by the decisions of the Science Planning Committee, the Academy remained the pre-eminent research institution. In addition, however, by the end of the period, efforts were being made to foster research in institutions of higher education and in ministries and other organizations concerned with production. The system enjoyed high levels of financial support, seriously lacked enough senior researchers, but was committed to reorienting the educational system toward producing large numbers of scientists and engineers, and was guided conceptually by the plan. The system was not without problems. This chapter will explore some of the more important problems and will then discuss some alternative models for science programs that developed in conjunction with the major shifts in national policy beginning in late 1957.[1]

Competing Models: 1949-1957

Chinese science policy in the first period was not limited to establishing formal organizations. Although the Chinese leaders had no experience with administering research prior to 1949, they would know from long experience that to make organizations work a set of mental and emotional orientations consistent with organizational procedures and goals must be present in the organization's members. The Chinese approached this requisite through a process of "ideological remolding."

In 1949, China had a pitifully small contingent of scientists and technologists.[2] The great majority had been trained in the capitalist countries, and the 1000 or so scientists at the Ph.D. level came from research traditions that were markedly different from that of the Soviet Union and from that which China wished to establish. Accordingly, the Party considered them "bourgeois" both in class background and in those acquired norms and behavior patterns that related to research. These included a sense of belonging to a community of scientists, of participating in that community through the publication of papers reporting research results, a respect for expertise in the resolution of technical problems, and the enjoyment of substantial autonomy—warranted by their expertise—in choosing and conducting research projects. This tradition placed great value on the judgment made by the community on the work of the researcher, and resulted in the motivation of the researcher being directed toward the achievement of colleague recognition. This set of orientations, which Western observers would associate with professionalism, became an object to be attacked by ideological remolding. In this first period, ideological remolding was relatively mild. Scientists would have to be converted to a socialist world viewpoint, but this conversion could take place naturally in the course of their work, as a result of research into the materialist basis of nature.

Thus, the foremost administrative problems in the science system were of two types. First, there was what appeared to be a technical problem of achieving coordination among the different research sectors, and overall policy control consistent with the plan. The second was a more subtle problem of integrating the science and engineering personnel, with their professional (or "bourgeois") orientations, into a system that had become hierarchically structured, and functionally specialized, with procedures governed by organizational regulations and short- and long-term plans. These became major tasks for the newly created SPC.

The issue of coordination—to achieve the kinds of intersectoral linkages described in the previous chapter—was particularly pressing. The nature of the research called for in the plan, and by science policy generally, was necessarily project-oriented and often interdisciplinary. In late 1956 the SPC, as an administrative expedient, divided all Chinese research into 26 task areas.[3] The enforcement of the plan required the disruption of the territorial boundaries of academic disciplines and of competing bureaucracies. An example of the coordination problem faced by SPC can be seen in the metallurgy field. The problem was to avoid duplication of research between the Institute of Chemical Metallurgy of CAS, and the Iron and Steel Research Institute of the Ministry of Metallurgical Industry. Both of these units were located in the western suburbs of Peking and both were doing research on oxidation in steel making. By a successful division of labor and tasks, it was said that some 850,000 *yuan* ($347,000) was saved in equipment costs, wages and maintenance.[4]

More complicated cases could be cited. One, in the field of hydraulic engineering, involved overlapping and duplication among the Hydraulics Research Department of the Ministry of Water Conservancy; the Hydraulic Engineering Research Office, run jointly by CAS and Tsinghua University; the

Department of Hydraulics at Tsinghua; the Hydro-Electric Power Research Department of the Ministry of the Electric Power Industry and the Water-borne Transport Research Department of the Ministry of Communications.[5]

It was stated explicitly that these reorganizations were not accomplished without controversy. Indeed, the controversy began while the plan was being drawn up.[6] The inability to resolve this controversy within the organizational setting that existed prior to the establishment of the SPC was precisely the reason SPC was required. The SPC then became the authoritative, final arbiter whose position transcended the intersectoral competition.

Other issues faced the SPC as well. In the liberalized atmosphere of 1956, and particularly the spring of 1957, there was fairly free and open discussion of the issues facing science administrators. Forums sponsored by CAS and the All-China Federation of Scientific Societies, which were reported in some detail, offered the outsider particularly good opportunities for viewing the stresses and strains within the science system. The issues that arose in these discussions may be divided into three categories.

The first were those that faced most, if not all, scientists. These included complaints that there was insufficient time for research due to political meetings and administrative responsibilities, that excessive secrecy frustrated scientific communication, that various aspects of the scientific information services (libraries, publications, distribution of foreign journals) were not working properly, and that the systems of research probationers and post-graduates, which had been initiated in 1955, were full of problems. This first category of issues also included criticisms of the blind application of things Soviet, and dissatisfaction with relations with Party cadres.

The second category involved selected individuals or groups of scientists, and many of these issues were related to the adverse effects that the introduction of planning had on certain fields. A microbiologist, for instance, complained that he was suddenly ordered to terminate a long-standing research project, and that he was not consulted about the sections of the twelve-year plan that affected his field.[7] Others complained that their suggestions were ignored. The geneticists had special complaints about "doctrinairism," which for a while restricted the publication of anything that was critical of Lysenko.[8] Certain geologists had complaints about the quality of geological surveying conducted by the Ministry of Geology.[9] Other complaints in this second category pertained to unfair practices in the publication of articles and in the system of academic awards set up in 1955, which was considered premature and an unwise use of funds.[10]

A final category of issues that came up for discussion concerned the role of CAS in the science system. Some complaints pertained to the way the Academy was run. Some scientists thought that "in fame for bureaucratism, the Academy of Sciences ranked second, if not first, among all the academic establishments.[11] This charge was particularly applicable to the post-graduate system and other personnel matters.

Another complaint concerned the extremely poor relations that had developed between CAS and the Ministry of Education. The problem was related to the bureaucratism in personnel management in both organizations (and in universities) and probably also to the uncertain involvement of universities in research at that time. One criticism of CAS from within its own ranks blamed the intersectoral conflicts on the exclusiveness of CAS:

Those outside the Academy of Sciences think that the entrance to the Academy is very small, there being a closed door on monopolistic ideas in the Academy. On the other hand there is inequitable division of labor and pay between those within the Academy and those outside of it. This deserves some careful examination by our comrades within the Academy. The walls separating the Academy from the outside world must be demolished, and so must be the internal partition walls that seem to exist between the various departments.[12]

As this sampling of the literature indicates, there were major contradictions within the science system toward the end of the first period. These contradictions can be understood in terms of the clash of conflicting models of science. The Chinese scientific community, which previously had understood the organization and management of research in terms of the primacy of professionalism, was being asked to submit itself to the bureaucratization of Chinese science. Table 4-1 contrasts the "bureaucratic model" with the "professional model."[13]

The decisions reached by the SPC at its fourth enlarged meeting in June 1957, immediately after the widespread debates that had occurred earlier that year, indicate that the views of the scientists were heard and some were acted upon.[14] The airing of views appeared to have a useful effect in achieving some accommodation between the professional and bureaucratic models. It appeared at the time that the chaotic situation that had arisen was brought under control and that the emergence of a new model—the "bureaucratic-professional"—was about to be realized. However, significant changes in national development strategy began to occur during the second half of 1957, which had the effect of prolonging the organizational uncertainty within science.

The Period of the Great Leap Forward: 1957-1961

The organization-building efforts for science in the 1949-1957 period were part of a larger national development strategy that like the science programs was strongly modeled after the Soviet Union. A few characteristics of this larger strategy are relevant. A first characteristic concerns the ordering of economic priorities and the organization of the economy. It is well known that the highest priority of the First Five-Year Plan (1953-1957) was heavy industry. Control over the economy was highly centralized in government agencies for production,

Table 4-1
The Professional and Bureaucratic Models for the Organization of Science

Organizational and Administrative Factors	Professional Model	Bureaucratic Model
1. Advocates	Scientists	Party and administrative cadres
2. Leadership and Party Influence	Leadership belongs to scientists of proven ability; grudging acceptance of administrative leadership in providing the material conditions for research, Party influence generally opposed except to the extent that it can facilitate the supply of the material conditions for research.	Primacy of leadership by non-scientists; separation of administrative and Party leadership in principle, but with recognition of the ultimate authority and primacy of Party leadership.
3. Centralization	Decentralized to level of experienced individual scientist.	Policy and administrative leadership centralized in SPC: academic leadership in CAS.
4. Differentiation	Favored, but on basis of skill (e.g., applied researchers should do applied research, basic researchers should do basic research regardless of place of employment).	Favored, but on the basis of the specialization of functions within a network of organizations (e.g., CAS as locus of basic research).
5. Recruitment	Controlled by scientists and based on demonstrated competence and ability.	Controlled by administrators/ Party and based upon organizational needs and procedures.
6. Personnel Mobility	Free and subject only to market considerations and individual interests and aspirations.	Administratively controlled and based upon organizational needs and procedures.
7. Rewards and Incentives	Although adequate compensation is expected, most meaningful rewards come from colleague recognition from the international scientific community; incentives, therefore to achieve that recognition through publication; some desire to serve the nation also present but subordinate to concern for recognition.	High salaries provided indicating high value regime places on scientists; state sponsored awards for achievement; efforts to undermine professionally relevant rewards.
8. Project Selection	Function of the internal development of the science; influence of the "paradigm."	Function of the needs of the state as outlined in the plan.

Table 4-1 (cont.)

Organizational and Administrative Factors	Professional Model	Bureaucratic Model
9. Project Control	Flexibility exercised by competent professionals.	Exercised by administrative cadre in consultation with professional.
10. Project Evaluation	Performed by leading scientists and ultimately by larger scientific community.	Performed administratively in reference to plan.
11. Communication	Informal, through journals and professional conferences; opposed to secrecy.	Formalized according to organizational procedures; secrecy justified in the interest of organizational integrity and national security; specialized mechanisms for information transfer and control established.
12. Foreign Collaboration	Support for principle of non-political internationalism in science; advanced training in capitalist countries experienced by many results in *de facto* orientation to literature from the West.	Internationalism in science linked to politics; "lean to one side" policy results in orientation to Soviet and Eastern European science; active attempts to discredit Western science.
13. Generational Relations	In principle, expertise and achievement, not generational differences, are bases for status differential; however, influence of traditional culture probably results in attitude that younger men must respect and learn from older men.	Older men must pay particular attention to training younger men; youth should learn science from older men, but efforts must be made to prevent older men from instilling "bourgeois" values in youth.
14. Extra-Academic Meetings	Political meetings opposed as useless; grudging and reluctant participation in administrative meetings.	Political meetings regarded as necessary, but limited to one day/week; great attention to administrative meetings and many experienced scientists coopted as administrators.

commerce, finance, and planning. Due to the availability of industrial technology from the Communist bloc, China's technological advance—as measured by the high priority sectors of the economy—showed a strong bias toward capital intensity.

China's emulation of the Soviet Union extended to the field of management. Managers were obviously chosen for their technical expertise, and they enjoyed considerable authority in the running of enterprises. They were subject to a chain of command leading up to the ministries, and this ministry-enterprise system in turn was subject to the requirements of the plan. Rewards to management were a function of plan fulfillment. This reward structure, coupled with authority patterns, created a definite, and by no means favorable, climate for innovation at the enterprise level. Technical innovation resulting from the importation and installation of industrial technology from the bloc countries was readily supported by managers since it was both consistent with policy and offered quick payoffs. The structure of managerial incentives, however, was not conducive to indigenous technical innovations.

Indigenous technical innovation in China during this first period, as we shall see in the following chapter, can be thought of as including "top down" innovation by trained engineers and designers, and "bottom up" innovation by workers spurred on to innovative activities by mass innovation campaigns. The frustration of both types of indigenous innovation, under the organizational conditions of the 1949-1957 period, violated China's spirit of nationalism, and the frustration of worker innovations also violated a basic tenet of Party doctrine on science and technology.[15] In a broader comparative sense, the drift of China's innovation experience was carrying her toward the technological dependence on industrialized nations that is so characteristic of Third World nations. Thus, when China initiated her first major policy shift in late 1957, changes in economic policy and structure spread into management and reward structures, choice of technology, the status of indigenous innovations and innovative capability, and of course the organization and operation of research and development.

In conjunction with the Third Plenum of the Eighth Central Committee of the Chinese Communist Party, the last quarter of 1957 saw a new expression of concern for the role that agriculture should play in China's development and also for a major decentralization of governmental activity.[16] In November 1957, Nieh Jung-chen, by then the chairman of the SPC, made an important speech on agricultural science. It intimated an expansion of regional research establishments for agriculture, gave guidelines for the relationship of these with the national Academy of Agricultural Sciences, which had been founded earlier that year, criticized scientists who resisted exposure to the production milieu, called for heightened ideological struggle, and emphasized the importance of coordination in research and the avoidance of duplication.[17]

By February 1958, CAS was preparing for a "great leap forward" in research. At a meeting of CAS institute directors, Academy President Kuo Mo-jo noted that the time was right for Chinese science to blossom—the twelve-year plan was drawn up, and although still a draft, it had been reviewed by the Soviet Academy. The SPC had been made a permanent state body, and new agreements for assistance from, and cooperation with, the USSR had been concluded. Attention was drawn to ideological remolding and the meeting concluded that the ideological factor was the constraint to be attacked. Chang Ching-fu, a

Vice-President and Party Secretary of CAS, urged the institute directors to assume the functions of combat commanders.[18]

The outlines for the entire science system for the "Great Leap" period were drawn up at the fifth meeting of the SPC in March. Although the "scientific choices" affecting established research in CAS probably continued to reflect the priorities of the plan, there was a decided shift in tone and emphasis that foretold the major institutional innovations that were to come. The outcomes of this meeting included a call for Chinese scientists to learn from the masses and from Chinese tradition as well as from foreign science, for scientists to adopt the "more, faster, better, and more economical" style of work, for the older generation to give more attention to training a new generation, and for all to engage in physical labor relating to their fields. With the renewed call for linking research with production, a decentralized science system was announced that would establish research activities as low as the county level. It was affirmed that non-expert administrative personnel could lead experts, but that these non-experts (often cadres who had been steeled in earlier revolutionary struggle) should make strenuous efforts to learn science and become experts in their departments. The SPC strongly encouraged more university research but urged that this be linked to production. Finally, the issue of ideological remolding was again raised, the conservatism of the scientists was attacked, and it was pointed out that some scientists were attempting to circumvent the plan by giving lip service to it while actually continuing their own research.[19]

The ideological problems of scientists were discussed in detail during this period by key Party representatives with science administration responsibilities. Chang Ching-fu, CAS' Party Secretary, noted three ideological problems that had to be solved. The first concerned the establishment of a proletarian perspective on scientific work and the liquidation of the bourgeois perspective. Under this theme, Chang criticized those who would "travel the socialist road in politics, but the capitalist road in profession," and who say "I listen to you in politics, you listen to me in professional work." Such attitudes, according to Chang, made it impossible to unite national needs with personal specialties for planned work. The persistence of the bourgeois perspective, according to Chang, also perpetuated such alleged abuses by the scientists as concentrating on personal fame and wealth; choosing irrelevant topics; using cover projects to hide actual work; allowing indecision, worries and doubts to result in switches in research topics; opposing collective work; allowing jealousies over "personal spheres" of research to interfere with work; resisting the criticism of others; stealing the labor of others; suppressing new talent and obstructing the advancement of others.

The second general problem concerned the need to abandon "institutional" concepts, and to establish "mass" concepts. The continuation of institutional concepts of scientific work was seen by Chang as resulting in the mystification of science, excessive specialization, and the stress on being "qualified" as a prerequisite for doing science. All of these, Chang argued, kept science from the masses.

Finally, Chang attacked the inferiority question noted in Chapter 2. He urged scientists to "solve problems independently, with confidence, and to eliminate reliance on others, particularly England and the United States."[20]

In another authoritative and relatively comprehensive statement on the ideological problems of the scientists, Fan Ch'ang-chiang, a leading non-scientist research administrator and policy maker with the SPC, linked ideological deficiencies to the basic social structure, norms, rewards, and motivations of the Chinese scientific community. China's scientists had been urged to sacrifice their sense of individualism and work for collective purposes. Individualism entailed autonomy in the selection of research topics and in conducting research, a desire to publish papers reporting research results and the enjoyment of receiving recognition from their colleagues. The scientists had apparently argued that to give this up was to undermine the motivation of the professional scientist. Fan set out to refute this argument by pointing out the adverse consequences of this "bourgeois ideology." He noted:

> Some scientific research personnel are not willing to exert all out efforts to solve the major problems affecting production today, but will often seek certain small issues which have no bearing on the general situation, and which other people have not taken up, in the hope of creating a name for themselves.

Fan noted as an example that some entomologists have ignored the problem of large scale destruction of crops caused by pests and have focused instead on rare species of insects for study.

> In the fields of botany and zoology, there are people who specially seek rare plants and animals, so that once they make a discovery, they can publish papers and furthermore name the species after themselves so that they may become world famous. As to what bearing these rare plants and animals can have on production, they are not concerned.

Fan admitted that in the early stages of capitalism this bourgeois ideology had a progressive role to play in combatting "feudal ideology." However, in a socialist state, bourgeois ideology not only had the effect of promoting irrelevant topics, it also frustrated team efforts required in complex projects, resulted in the disregarding of worker and peasant innovations, and perpetuated undesirable social stratification which separated the intellectual from the masses.

The problem of motivation thus was to achieve a correct political and ideological stand. Fan urged the scientists to look to the masses, to observe the "motive power" (*tung li*) they were displaying, and to learn from and emulate them. Fan noted that it was particularly important that the younger scientists have the right ideology, and not "linger after fame and profit." Instead, they were urged "to serve the people and solve problems for the state."[21] Thus, Fan and others were setting the goals for ideological remolding, goals that were to be realized through political meetings and campaigns featuring ideological study, criticism, and self-criticism by the scientists.

The objective of this "administration by political campaign," aimed at correcting ideology, can be seen in the following evaluation of the effects of the rectification campaign of 1957-58 by SPC chairman Nieh Jung-Chen:

> The rectification movement among scientists will produce far reaching effects on the promotion of science. . . . The Experimental Biological Institute of Shanghai is a case in point. In the past it has been the practice in the Institute to establish sections or divisions according to the scientists available. Almost every high ranking research scientist led a section or division, and set up his own personal domain in his own specialized field. As a result, forces were divided and the sections or divisions did not cooperate with one another. The rectification movement has put in good order the thinking of the scientists and persuaded them to acquiesce in a suggestion for reorganizing all the personnel of the institute into four sections or divisions in accordance with the four key problems being tackled, thus, settling the question of internal cooperation.

This report goes on to note that rectification also had a beneficial effect on external cooperation with other units as well.[22]

Complementing the ideological emphasis of the summer of 1958 were a number of organizational developments that were characteristic of the emerging new model. A concerted effort was made to bring established science into direct contact with the reported upsurge of worker and peasant "science," by convening conferences of scientists, technicians, workers and peasants, by admitting workers and peasants into research institutes and professional societies, and by extending the work of the societies to low levels of the social order.[23] The operative organizational term for this period is "scientific network." Parts of the network were established during the 1949-1957 period as we have seen, but beginning in mid-1958 there was a proliferation of new organizations–CAS branch academies, science committees at provincial and subprovincial levels (down to the county level), associations of science and technology paralleling the science committees and devoted to popularization, and "scientific research groups" in factories and villages. Part-time technical schools for workers and peasants were set up at different administrative levels.

To coordinate the work of popularization at the national level with the capability of the trained scientists, the Science and Technology Association was established in September 1958 by merging the All-China Federation of Natural Science Societies and the All-China Association for the Dissemination of Scientific and Technological Knowledge. As we have seen, to achieve better administrative and policy control over the emerging "networks," the SPC merged in November with the State Technological Commission to form the State Scientific and Technological Commission.

Although the themes of this period emphasize ideological remolding, decentralization, and mass science, support for established science increased. To meet the need for highly trained personnel, the University of Science and Technology was established jointly by CAS and the Ministry of Higher Education in November. Expenditures for science continued to increase, as did the number of institutes under CAS.[24] The overall strategy was one of "walking on two legs,"[25] although the changes during this period did mean that the leg of professional science should develop a fundamentally different *modus operandi.* This shift can be discerned in the emergence of the mass movement, or mobilization style in the management of research at the institute level.

The mobilization style attempted to break down the conventional patterns of scientific work and substitute instead "forums of humble discussion" and a technique known as "probing from all sides." The former was characterized by the convening of extraordinary meetings to discuss problems encountered in research. The objective was to take the problem out of the hands of the specialist who had been working on it and lay it before other scientists, technicians and workers for discussion. The "probing from all sides" technique involved attacking a problem from a number of different approaches, choosing the more promising ones, and then finally concentrating all resources on the most hopeful.

A number of advantages were claimed for the mobilization style. It had the advantage of bringing people together from different branches of science to focus on a new problem. This was important because it was felt that:

> . . . the starting point of the growth of modern science is usually found at an intersection where several branches of science converge. Thus, it is rather difficult to overcome difficulties in research work by relying only on people engaged in any one branch of science. Mass movements in scientific and technological work can pool together the wisdom of people from all circles to get a speedy breakthrough.[26]

It was also felt that the mobilization style would bring together theory and practice by bringing workers and peasants into contact with scientists. This was considered desirable because it was thought that workers and peasants are often the first to spot the applicability of new knowledge for production. Furthermore, . . . an experiment cannot be considered completed until it has succeeded in experiments made in actual production.[27] The following case is illustrative of the mobilization style.

The Institute of Applied Physics, CAS, had apparently been assigned the task of research on semi-conductors. The senior scientists, who occupied positions of academic leadership and who were allegedly under the influence of the bourgeois ideology, reportedly were cautious and conservative in their research procedures and were skeptical of Chinese indigenous capabilities and resisted learning from the Soviet Union because of a worshipful orientation to Western science. In keeping with the times, a series of "leap forward" conferences were convened under Party initiative. Out of one of these conferences came a work target to develop a transistor that would be up to world standards. This was to be done quickly, by a "sneak attack." The laboratory that was to take up this challenge began by convening a meeting of all personnel. At the meeting, the target and, importantly, the political significance of the target were discussed. The younger scientists then convened their own meeting, discussed their position, and prepared verbal attacks on their superiors. Another full meeting was convened; the elders who were criticized accepted the criticism, and all were reported to have shared in increased political awareness. As the work began, four different approaches were tried, but when one approach began to yield results, all efforts were shifted to it. Once the transistor was produced, it had to be tested, and this required assistance from another group. Resistance was encountered there, however, from a U.S. trained scientist. The younger people persisted, and allegedly did not allow the "conservatism" of the "bourgeois" expert to stand in their way.[28]

The case indicates a style of work under Party leadership, where meetings for criticism and self-criticism are convened, resources are mobilized, enthusiasm is generated, and the conservatism of the older experts is overcome. The generational cleavage is characteristic of the period. During this time the elders were criticized not only for conservatism, but also for treating the younger men as personal property. They were accused of remaining aloof from subordinates and of refusing to explain directives to, and discuss technical problems with them. Unfortunately, it is impossible to corroborate this charge but reports of the frustrations experienced by younger scientists in other low-income countries suggest that it is not altogether implausible.[29]

The model that emerged during the "Great Leap" years (1957-1961)–the mobilization model–is based on a rather literal implementation of the views of scientific development and the role of science in society that are contained in CCP ideology. These views include the principles that: (1) scientific development is a function of social and economic factors external to science itself; (2) knowledge is the product of the unity of theory and practice; (3) science must serve production; (4) the ultimate source of discovery and verification in science rests with the laboring masses; (5) science can be, and must be led by the CCP, the representatives of the masses; and (6) the establishment of a unique Chinese scientific tradition rests with the younger generation of scientists who must be prepared to rebel against the older generation if necessary. The mobilization model is summarized in Table 4-2.[30]

The policies of the "Great Leap Forward" had the effect of reorienting the trends in the economy mentioned above. The policy of decentralization was particularly important but requires some qualification. The decentralization that occurred was to the level of provincial control, although unlike the decentralization in the Soviet Union at the same time, the central ministries were not

Table 4-2
The Mobilization Model for the Organization of Science

Organizational and Administrative Factors	Mobilization Model
1. Advocates	"Maoist" or "radical" elements in the CCP.
2. Leadership and Party Influence	Principle of Party leadership literally and rigidly enforced; Party leadership extends into academic matters and influence of scientists in decision making sharply curtailed.
3. Centralization	Simultaneous centralization and decentralization. Policy-making authority, and areas of more sophisticated research highly centralized under SSTC and CAS and central ministries respectively; nevertheless, growth of a regionalized research network including provincial branches of CAS and "research groups" attached to production levels; regionalized network under administrative supervision of provincial Party Committees.
4. Differentiation	Under attack; scientists are to become workers, workers are to become scientists; organizational boundaries are attacked as impediments to release of mass initiative; factories and farms become not only units of production, but units of education and research as well; nevertheless, some differentiation maintained as the "two legs" strategy implies and as establishment of new institutions like University of Science and Technology indicates.
5. Recruitment	Controlled by Party and based upon political and ideological considerations.
6. Personnel Mobility	Controlled by Party and based upon political and ideological considerations.
7. Rewards and Incentives	Although there is no indication that high salaries for scientists have been abolished, they are greatly deemphasized; the reward system of the scientific community (i.e., colleague recognition) is intensely criticized; state sponsored awards for achievement suspect; non-material, ideological incentives emphasized.
8. Project Selection and Goals for Research	Principle of science serving production emphasized; as organizational barriers break down and scientists are sent into productive milieu; they are, with cooperation of the masses, to choose topics and apply their knowledge and abilities in response to production needs; plan loses significance in that it is constantly revised[a] and in this sense work becomes less routinized.
9. Project Control	Exercised by Party; although lip service is paid to relevance of plan, ideological criteria predominate.
10. Project Evaluation	Performed by Party in reference to ideological and immediate production criteria.
11. Communication	Attacks on stratification, differentiation and organizational boundaries intended to open up communications within the system; however, the mandating of communication and the intensity of campaigns may result in psychological barriers to communication; secrecy continues.

Table 4-2 (cont.)

Organizational and Administrative Factors	Mobilization Model
12. Foreign Collaboration	Greatly de-emphasized as native methods and native wisdom are celebrated; "cultural nationalism" to the fore.[b]
13. Generational Relations	Generational differences exacerbated; youthful frustrations channeled into criticisms of elders as youthful rebelliousness is legitimized; distinction between older men as sources of scientific knowledge and as role models blurred.
14. Extra-Academic Meeting	Indiscriminate political and administrative meetings disrupt research routines and cut into time available for research and teaching.

[a]See Mikhail A. Klochko, *Soviet Scientist in Red China* (trans. Andrew MacAndrew), (New York: Frederick A. Praeger, 1963), pp. 95-97.

[b]For a discussion of the concept of cultural nationalism as it applies to medicine, see Ralph C. Crozier, *Traditional Medicine in Modern China: Science, Nationalism, and the Tensions of Culutral Change* (Cambridge: Harvard University Press, 1968), p. 99ff.

abolished. As one observer has noted, "Probably at least 40 to 50 percent of the whole industrial sector remained in the ministerial system after decentralization" (as opposed to 60 to 70 percent during the initial period).[31] It was chiefly light industry, and importantly, commerce and finance that were subject to decentralization. As a result, provincial and subprovincial authorities had new sources of funds and increased freedom of investment.

The increased reliance on non-material incentives, coupled with decentralization, had the effect of politicizing the economy. The position of the Communist Party, particularly at the provincial and enterprise level, was greatly enhanced. The party became the chief instrument for manipulating non-material incentives and for performing the all-important coordinating function in the absence of central financial and commercial controls.[32]

The increased role of the Party at the enterprise level effected changes in management. The "one-man" management system of the First Five-Year Plan period gave way to more collective arrangements with party cadres taking the lead, to be sure, but with worker participation.[33] The reward structure for workers built around piece work, which spread during the First Five-Year Plan, "was cut back and abolished on a wide scale."[34] The new management apparatus was still subject to production targets. However, the absence of effective control mechanisms, the presence of policy *cum* ideological injunctions to support worker and peasant innovations and the primacy of Party cadres in enterprises, who presumably were responsive to ideology, created a rather different climate for indigenous innovations than previously.

Another change in economic strategy during the "Great Leap" years involved choice of technique decisions. One of the key reasons for the policy shifts of

1957 was that China's vast labor force was being inadequately employed. Part of the strategy for change, associated with the organization of communes, was to use surplus labor on public works and irrigation projects. Another dimension of the strategy, however, was to give far higher priority than previously to the growth of small- and medium-sized industry that would use labor intensive techniques.[35]

Finally, the situation in agriculture in 1957 called for major policy changes. Anthony Tang has nicely summed up the choices facing the leadership at that time.

> The traditional sources of agricultural growth appeared to be drying up. Moreover, agriculture continued to limit the rate of industrialization. The choice was between (1) pressing ahead in agriculture along the traditional margins through drastic reorganization of production and household activities, or (2) modernizing the agricultural sector by instituting thoughtful research, development, and information programs.[36]

The Chinese chose the first option, although efforts were made to bring science and technology more to bear on agriculture than previously.

The "Great Leap" strategy, as implemented in the 1957-1960 period, failed. The decentralization actions had the effect of making resources available for new economic initiatives but failed to provide for effective coordination. The results were supply and delivery problems as well as the establishment of new industry, much of which turned out to be ill-conceived and uneconomic. Decisions for agriculture were manifested in the formation of communes that drastically disrupted the ownership and work patterns of rural China.[37] These organizational difficulties, coupled with bad weather conditions, resulted in serious crop failures and a major retrenchment from the "Great Leap" strategy.

In spite of the failures, the "Great Leap" economic programs, along with the science programs, represented a significantly different vision for Chinese society. The economic, science, social, and political programs were intended to go together, and represented an alternative model for Chinese development.

The Period of Consolidation and Rationalization: 1961-1966

The search for a new approach in science during 1957-1961 was closely related to the "Great Leap Forward." Both were radical experiments in social organization for which there was inadequate preparation, implementation and anticipation of consequences. As a result, the serious disruptions in Chinese life that occurred were not channeled into productive modes but instead had the effect of contributing to social disintegration. The Party was, therefore, forced to institute policies of retrenchment and readjust its leadership style.

The economic difficulties associated with the "Great Leap" are presumed to have had the political consequences of reducing Mao Tse-tung's authority in national policy making and enhancing the authority of Liu Shao-ch'i and other regulars in the central Party apparatus. The policy measures of the early 1960s are often attributed to Liu and his followers, although it also appears that Mao Tse-tung was planning a comeback, with the help of Lin Piao, by developing his own network of influence within the People's Liberation Army. By 1964, it appears that Mao's weight was again being felt.

The economic retrenchment that began in 1961 was characterized by a shift in the locus of economic decision making and a commitment to a more "developmental," as opposed to "extractive," approach to agriculture.[38] The local industry strategy of the "Great Leap" was not abandoned but was subject to rationalization and was made to focus more directly on the needs of agriculture.

The reorganization of the economy in 1961, which represented a decentralization of a different kind, was coupled with a recentralization of certain key coordination functions. The power of production units was greatly enhanced as managers again came to enjoy "independent operational authority," while authority over finance, trade, and communications reverted to the center.[39] In agriculture, there was a major retreat from communization. Although communes continued to exist, and the concept of "three levels of ownership"—communes, production brigades, and production teams—persisted, the basic accounting unit again became the production team, which corresponded to the natural village.[40]

The widening of the Sino-Soviet rift and the eventual suspension of Soviet assistance forced major readjustments in the Chinese economy. The emphasis became one of self-sufficiency, particularly in such fields as producer's goods for agriculture, chemicals, and petroleum.[41] Self-sufficiency in the early part of the 1960s meant self-sufficiency for the nation, not necessarily for the regions. Thus, in agricultural development, the initiative came from the center, and entailed the funneling of centrally supplied inputs to selective high yield areas.[42] Similarly, in agricultural mechanization, the shift from the "Great Leap" strategy meant an emphasis on technical rather than social reform and a change in the locus of control over mechanization from local levels to the center.[43]

In retrospect, it appears that after the initial "post-Leap" recovery period, the Liuist leadership set out on an industrial development course that had significant implications for the allocation of centrally controlled resources and for the character and direction of science-related activities. Liu apparently favored, and is alleged to have promoted, electronics over iron and steel as the basic focus for Chinese industrialization.[44] The former is a research-intensive industry that would take China directly into the electronic age, through the advances of modern science. A related issue was the Liuist program for the standardization, specialization, and verticalization of certain key industries. Standardization involved the establishment of national standards for industrial production with the elimination of qualitative verification.[45] Such standardization again involved the enhancement of the position of the expert and the position of the central industrial ministries. Having the same effect was the establishment of what the Maoists called "trusts" or specialized industrial sectors under vertical control from the center, free from regional political and administrative influence.[46] The implications for the science system included the relegation of the decentralized, participatory research activities of the "Great Leap" period to a low-priority position and their replacement by organizationally distinct, sophisticated, project-oriented research subordinate to the center.

Clear statements of research priorities in the 1960s are difficult to find. Some inferences can be drawn, however. First, the persistent agricultural dilemma and the more immediate failures of the last years of the "Great Leap" resulted in the intensification of agriculturally related research and of efforts to link research to a delivery system organized around demonstration farms.[47]

Second, the withdrawal of Soviet technical assistance meant that China's prime source of new technology was lost. Although China did turn to non-Communist countries for new technology in the early 1960s, it was also necessary to redirect her indigenous R&D, at least in some areas, toward the generation of technologies that were to have come from the Soviet Union.

In general, however, it seems likely that the research priorities set during the period of organization-building continued into the 1960s, with some adjustments for agriculture and for the withdrawal of Soviet assistance. The twelve-year plan was to apply until 1967, and there is no reason to think that it did not. However, with the abandonment of the mobilization model, and the new prestige accorded to expertise in the 1961-1966 period, professional scientists and engineers now had a considerably greater voice in translating the long-term objectives of the twelve-year plan into short-term operational plans.

By 1961, it was clear that a major shift away from the mobilization model was occurring. In the summer of that year, reports appeared that administrative regulations for research establishments were being revised to provide a better professional environment for research. Foremost among these was the rationalization of the researcher's time, with a return to a guaranteed 5/6 of a week devoted to academic matters. In addition, emphasis was placed on strengthening the administrative structures to insure that the material conditions for research—instruments, publications, reagents—were satisfactory and to facilitate the planning and reporting of research.[48] These changes marked a major shift from the mobilization model back to the bureaucratic-professional model that had begun to take shape in 1957. This model was characterized by a reduction of the influence of ideology and politics, the orderly conduct of administrative business by a committee for institute affairs, and the assignment of substantive matters of research to an academic committee led by scientists (see Table 4-3).

The professional side of the bureaucratic-professional style is characterized by the prominence of a research leader, such as a laboratory head, whose position and authority are a function of his expertise and experience. He is usually older than his subordinates. The style of his leadership depends very much on his personality. If he is competent and dynamic, he may encourage his subordinates to master the existing paradigm, encourage participation on the formulation of plans and procedures, and encourage free and open academic discussions. The research leader is knowledgeable about the research work going on below him, keeps firm but flexible control of it, is accessible to subordinates, is actively involved in research himself, and serves as a creative role model for the younger men who are in the process of gaining experience in research.[49] In the bureaucratic-professional style, the research environment is more ordered, specialized, and stratified. Assuming leadership competence on the part of laboratory heads and administrative and Party cadres, communications would be open but regularized by procedures. This is true both for internal communications and for communications with other research and production units in the environment.

Although there were major consolidations of the branch academies and the low-level scientific experiment groups were either downplayed, discouraged, or discontinued, the 1961-1965 period was unique in that there were few significant organizational changes in the conventional research segment of the science system. A few research installations were added in the various sectors, but the basic structure remained the same. Instead of further reorganization, attention turned towards making the system work. In the regime's perennial quest for effective ways of linking research with production, it urged open and widespread discussion of this topic.[50] This produced an outpouring of views about the nature of research, the differences among research, development, extension work and popularization, and the limits of responsibility for those involved with these various activities.[51] Did research end with a report or published paper, did it end with the building of a prototype, or did it end with a new product? Where did a researcher's responsibility end and a designer's begin? Should an agricultural scientist limit his activities to the laboratory, to an experimental field, or should he go among the peasants and go through an entire crop cycle with them?[52] A related issue was whether science could or should lead production, or whether production could or should lead science. Was science a part of the superstructure, or a direct productive force?[53]

The orientation of the Liu Shao-chi/Teng Hsiao-ping leadership during this time was to make accommodations with professionalism. Steps were taken to improve the educational system by providing longer and better training to those whose aptitude was demonstrated in conventional academic pursuits. Scenic vacations were provided for scientists and their families, and there was a reintroduction of a system of academic awards and material rewards for invention.[54] In keeping with this orientation, research activities could be separated from development, extension, popularization, and production by introducing an element of functional differentiation and specialization. The high hopes attached to scientific research may indicate that at least some key Chinese leaders began to see science as a direct production force.[55]

In addition to the preoccupation with linking research with production, there was a concern for rationalizing the research process itself. This subject was taken up in a major article in *Hung Ch'i (Red Flag)* on the organization and management of research by Ch'ien Hsueh-sen, one of China's most celebrated scientists. Ch'ien, an aerodynamicist who was persecuted by the U.S. Government in the early 1950s, returned to China in 1955 and had by this time become a party member and a leading member of the emerging Chinese scientific establishment. Ch'ien's statement can be taken as representative of the professional's view of the bureaucratic-professional model.

Ch'ien began by noting that the characteristics of modern research set it apart from ordinary production activities. Modern research was highly specialized and required elaborate equipment. In modern research, the time for bringing results of fundamental research into practical utilization had been shortened. However, the characteristics of modern research—in contrast to traditional science—required the resources of complex organizations in planning, in material support for equipment, supplies, construction and maintenance, and in providing information services. Ch'ien noted that research units must keep the "leading organs" (i.e., centralized administrative and policy-making bodies) informed through concise, accurate and opportune reports, but he also urged that modern science be employed in research administration itself. Ch'ien gave as examples the potential applicability of computerized data storage and retrieval, dynamic programming, and operations research.

The key to achieving success in the work of research organization and management (which in turn held the key to productive research) according to Ch'ien, was to train and employ specialized research managers. Ch'ien thought that initially these people must come from the ranks of the researchers but that in the longer run, specialized training for research managers should be encouraged.[56]

From Ch'ien's article and from other literature of the period, we can see that the emergence of the bureaucratic-professional model in China involved organizational processes that are not dissimilar to those found in the United States as science interacts with complex organizations. Kornhauser, for instance, in studying the interaction of science and individual firms found that conflicts occur over goals for research, control over research, incentives and influence but that these can be accommodated through various mechanisms.[57] Conflicts over goals, for instance, are accommodated by organizational differentiation. Basic research, applied research, and development become located in organizationally distinct settings. A similar process occurred in China for the science system as a whole as a result of the decisions of 1957, which were fully implemented after 1961.

Conflicts over control are accommodated by the emergence of the "scientist-administrator," who has both the competence and the sensitivity to deal with the professional community, while at the same time represents and is accountable to the formal organization. Kornhauser's scientist-administrator sounds very much like the role of research manager advocated by Ch'ien Hsueh-sen.

Conflicts over incentives are typically accommodated through a process by which the scientist is resocialized to be amenable to organizational rewards. Again, it would seem that by the early 1960s China's scientists were being subject to the prospects of attractive material rewards—high salaries, vacations, cash prizes—as well as to non-material honorary prizes given by the state,[58] in return for working for the goals of the state.

At the level of national research administration, the perennial problem of coordinating and linking together the entire system persisted. During the organization-building years, it was felt that the system of CAS departments and concurrent positions would be an appropriate mechanism for dealing with this task. We have seen the serious inadequacies of this system in the face of an expanding and ever more complex national science program. With the creation of the SPC, and eventually the SSTC, the role of the department system was greatly diminished in *national* policy making. The SSTC, however, by its very nature, required mechanisms for intersectoral control. It needed informational inputs for its decision making, beyond that provided by the competing sectoral research bureaucracies, and a means of orchestrating intersectoral cooperation. By the late 1950s, the SSTC seems to have realized the potential of using professional societies for this task, and by the early 1960s, the activities of existing professional societies were vigorously encouraged, and many new societies were organized.[59]

By 1962, it appears that the new formula for using professional societies as administrative mechanisms for coordination was being viewed with satisfaction. Leading departments, on the basis of the needs of production, were to raise problems for the societies to study and solve.[60] There was an implied recognition that a certain autonomy for the societies would enhance their usefulness, since "it is necessary for the control of academic activities to reflect the demands of the scientific and technical workers themselves."[61] By 1963, the new formula was being celebrated as having produced a bumper year of 55 academic conferences of the national societies, and it was explicitly recognized that "Societies of natural science are the best forms of organization for the different departments and units to get in touch with the specialists of the same trade."[62]

We can get an idea of the role of the societies in national policy making for science by piecing together information on the Astronomy Society from the early 1960s, and revelations made during the Cultural Revolution. In late 1961, the Astronomy Society sponsored three meetings in cooperation with the academic committee of the Tzuchinshan Observatory of CAS. Each was topically oriented: one on fixed stars and evolution, one on celestial mechanics, and one on solar physics and radio astronomy. The meetings were considered quite productive in summing up recent developments in the field, not only for putting forward "many constructive suggestions concerning the long-range planning of various branches of science," but also for working out "arrangements for the current work of existing astronomical institutions, the directions of cooperation between the astronomical and other circles, and the tasks of astronomers."[63]

The astronomy meetings, like those of other societies, were supposed to be the centers of "blooming and contending" with diverse views presented. One such dispute occurred at the solar physics meeting at which two papers, using the same data, written by two men from Tzuchinshan, reached opposite conclusions about the peak value in the cycle of solar activity with special reference to the 1967-68 period.

This case is interesting in light of news of Chinese research activity during the solar eclipse of 1968. According to these reports, the only reason that China did any research on the eclipse at all was because the Cultural Revolution had encouraged younger "revolutionary" scientists to break their bonds with the older, more conservative "authorities." These authorities had argued that given China's limited technical capabilities, the kind of data that could be obtained from mounting a major solar eclipse program could be in no way commensurate with the costs. Accordingly no plans had been made for such activities.

This basic decision went back to the 1962 Conference for Overall Planning of Scientific Research Work.[64] An account of the role scientific societies played in planning can be hypothesized from the fragmentary data available. The 1962 Overall Planning conference, as a selective gathering, was presumably convened by SSTC for setting national research priorities. The reports from recently held, society-sponsored academic meetings, which represented the most thorough summing-up of the state of a discipline, would represent major inputs of technical information into the SSTC decision-making process. It is rather implausible that in making a major decision affecting solar physics—the eclipse decision—that the results of the most recent technical meeting on the subject—the 1961 Astronomy Society meetings—would be ignored.

This case highlights also the significance of the planning process itself. Technical results, to the extent they are used in planning, take on political significance far more quickly than they do in a less-planned system. Planning also is a powerful channeling device, which as the solar eclipse illustrates, can set the course of and therefore the funding for research for years in advance. As such, it is a type of political resource within the science system that gives individuals or groups with access to its control considerable leverage. But the planning as hypothesized here also opens up the possibility for a kind of advocacy process in which an individual scientist or group of scientists can use the floor of society sponsored meetings to argue for a position, and perhaps change the direction of a field by winning a place in the plan for some new topic. Sitting as observers and judges of this process would be the non-expert state science administrators.

Apart from the concern for good research management as expressed in Ch'ien's article, and the larger problem of coordinating national research efforts, the 1961-1966 period also saw increased concern for the quantity, quality and utilization of scientific manpower. The manpower question had long been a significant one and the limited supply of senior scientists and technologists was one of the most important constraints on scientific development.[65] The relative scarcity of experienced manpower in turn made the question of utilization a particularly important one. Scientists had long complained that they were not being properly utilized, and it appears true that they had excessive administrative responsibilities, were often transferred from one job to another or from one unit to another, and were assigned to jobs for which they were unsuited.

A new approach to these questions appeared in 1964. On May 29, *People's Daily* editorialized on the need to reduce the arbitrariness and irrationality in the training and utilization of scientific manpower. One week later the sentiments of the editorial were given organizational expression in the establishment of a Science and Technology Cadres Bureau under the State Council. The Cadres Bureau appears to have been a centralized, specialized, super-personnel management agency for rationalizing the manpower problem through administrative means. The fact that this bureau was made an organ of the government, and not the Party, may indicate the degree to which this important but complex aspect of national science development was being depoliticized and regarded instead as a technical problem requiring administrative rationality.

These attempts at bringing greater effectiveness to the science system stand in sharp contrast to the approach employed during the "Great Leap." In the latter case, as shown in the previously cited statement by Nieh Jung-chen, organizational effectiveness was closely linked with ideological remolding. This type of approach would place greater emphasis on the role of the Party and its capacity to elicit changes in human behavior in accordance with ideology. The technical aspects of administration would be neglected for the sake of a series of processes that in some ways resemble what in the West would be called "organizational development." During the period of consolidation, however, improving the effectiveness of organizations was approached by the rationalization of administration. This type of approach would strengthen the hands of those in the Party and government who rejected the idea of administering science by mass movement with the result that "Party leadership" would be interpreted to mean administrative leadership and the character of the Party changes accordingly. The fact that the bureaucratization of the Party extended far beyond science and that Mao and his followers opposed this change was one of the central issues that led to the Cultural Revolution.

The Period of Radical Destruction and Radical Reconstruction: 1966-1971

Although the Cultural Revolution was not about science directly, many of the key issues involved had significant relationships to and implications for science and technology. Questions of national defense, educational policy, industrial and agricultural strategies, and the cultural transformation of the Chinese masses, all impinge upon science policy. At stake were competing visions of the future of Chinese society, including the relationship of science and technology to the distribution of power. Powerful elements in the Communist Party and the government bureaucracy had vested interests in the continuation of the society that had emerged during the early 1960s with its organizational and policy priorities. Such a society apparently was not consistent with what Mao thought was in the best interests of the masses. The Maoist strategy, therefore, was to stimulate revolution against established authority. The result was three years (1966 to 1969) of serious disruption in the science system, followed by three years (1969 to 1971) of reconstruction along Maoist lines.

One of the first signs of the Cultural Revolution in science occurred on July 30, 1966, when two important science administrators and party representatives—Han Kuang, Vice Chairman of SSTC, and Chang Ching-fu, a Vice-President of CAS and its Party Secretary—were criticized at a mass meeting by Wang Hsi-p'eng and Chang Pen.[66] The latter two persons, relatively unknown before this, experienced a dramatic rise in their prominence within CAS and SSTC respectively during the Cultural Revolution, and then like many others, disappeared from the limelight as the Cultural Revolution wound down. This meeting set the course for a year of relatively autonomous revolution within these two organs in which criticism of the administrative establishment developed from the bottom up.[67] Thus, when the important 11th Central Committee Plenum produced the sixteen-point Charter for the Cultural Revolution in August 1966, with its apparent guarantee against disruption in scientific research (point 12), the stage may have been set already for an "orderly revolution" within the science system.

In the course of the Cultural Revolution, however, the plans of 1966 proved to be inadequate. Although Chou En-lai claimed in mid-1967 that it had not been necessary to send special political teams into CAS during the first year of the Cultural Revolution,[68] by the end of 1967 it was clear that there had been major disruption in the Academy and that Mao Tse-tung Thought Propaganda Teams, under the leadership of the PLA had been dispatched to CAS. Similarly, during 1967 the SSTC came under sharp attack. Indeed, the SSTC was the science system's major casualty, for while CAS was disrupted and changed, it survived. SSTC did not. The members of the SSTC, including its chairman Nieh

Jung-chen, were attacked, and some of them disappeared from public life. Some have recently re-emerged but the Commission itself seems to have been eliminated. This fact tells us something about the impact of the Cultural Revolution on science; namely, that the fury of the Cultural Revolution was directed far more against the science administrators who made and carried out policy than against scientists themselves.[69]

The remainder of 1966 saw the appearances of reports reflecting themes from the 1957-1961 period. Examples of workers and peasant innovations were hailed and an account of the world's first synthetic benzene workshop indicated how mobilizational management had released the energies of younger researchers and technicians. Once the proper line had been established, these "revolutionaries" overcame the conservatism of a U.S. trained expert who tried to frustrate their efforts with counter experiments and by withholding funds and materials for the work.[70]

The year 1967 saw the unfolding of attacks directed explicitly against Liu Shao-ch'i and many science policy positions identified with him. At a meeting of representatives from "scientific and technical circles" in Peking in April, Chang Pen laid down a barrage of charges against Liu. The radical line of 1958 was hailed and Liu was accused of systematically undermining that line during the early 1960s. It was Liu, according to Chang, who gave up party leadership in science in favor of leadership by bourgeois intellectuals, and who promoted "high salaries," "high awards" and "high rewards for articles," despite a 1960 order from Mao that the giving of medals and conferring the title of "doctor" should not be done.[71] Similarly, in policy for higher education, Liu and his followers were accused of fostering elitism and of aping the Soviet Union in the "Regulations Governing the Award of Degrees, Academic Titles and Honorary Titles" and the "Regulations Governing the Training of Associate Doctors and Post-graduates," which they promulgated during the 1961-1965 period.[72] In addition, Liu was accused of "slavishly" seeking advanced technology from abroad with little concern for China's indigenous capabilities.[73]

By the second half of 1967, there had been enough chaos in all sectors of the society, that the People's Liberation Army was asked to intervene in a major way under the banner of the triple alliance–PLA, workers, and revolutionary cadres. A revolutionary committee was established at CAS during the summer, and during the remainder of the year, worker and PLA teams assumed leadership of other research organizations. Revolutionary committees were established throughout the government and under this rather uncertain administrative arrangement, major reforms in the structure of society were begun during the second half of 1968. These included the decentralization of the management of productive enterprises, increased worker participation in management, the simplification of administration, the rejection of material incentives, and the mass transfer of urban-educated youth to the countryside on a permanent basis. Related to the latter was the growth of May 7th Cadre Schools, where urban cadres and intellectuals were re-educated through labor, and either passed through these schools on the way to permanent reassignment in the countryside, or were expected to spend a few months there each year.[74] The principle of combining functionally differentiated activities was pushed by celebrating the experience of the Shanghai Machine Tools Plant as the appropriate model for fusing research, training, and production.[75]

The era since the Ninth Party Congress in April 1969 has shown a continuation of political instability, particularly at the center, and considerable organizational indeterminacy within the government. It appears that one consequence of the Cultural Revolution has been some devolution of power from the center. As a result, the relationships between Peking and the provinces have appeared uncertain. Much time and effort have gone into the restructuring of the Communist Party and the transfer of powers from the PLA back to the CCP.

The criticism of Liuist policies has continued since April 1969 as has the restructuring of society. A campaign to get rid of superfluous rules and regulations in the management of enterprises emerged during 1969 and 1970. The Liuist notion of "technicians run factories" has been sharply criticized as worker participation in the management of enterprises was encouraged.[76] The support of local industry, the restructuring and reduction of the central bureaucracy and the large-scale movement of urban personnel to the countryside provided the decentralization strategy of the late 1960s with a structurally different quality than that of the late 1950s. The vision that was being institutionalized was that of a highly decentralized society, committed to regional development in which much of science and education are locally administered and closely related to local industry and a local resource base.[77]

By 1972, not only were the Liuist deviations to the right criticized, but radical left positions associated with the fallen Lin Piao also came under attack. This has resulted in a retreat from many of the more radical practices and policies growing out of the immediate post-Cultural Revolution period. There has been widespread concern that the expertise of scientists and engineers be fully employed. It appears that in the enthusiasm to discredit the Liuist notion of technique in command, a leftist deviation set in which undercut the positions of technical personnel, and wasted their talents and which became a fitting subject for criticism. Similarly, the criticism of Liu's alleged "slavish comprador" attitude towards foreign science and technology produced a leftist anti-foreign-ism, which in turn was rectified by implementing the line of "make things foreign serve things Chinese."

In terms of the formal organization of science-related activities, it appears that the SSTC has been replaced by a Science and Education Group under the State Council. This group is headed by Liu Hsi-yao, a former vice chairman of SSTC. However, this new organization may not have the all-encompassing authority over science and technology that the SSTC once had. Some of the other changes in the organization of research help explain why this is so.

Many of the institutes formerly under CAS control apparently have been shifted out of the Academy. Although there is no definitive information on this point, it seems likely that many of these would have come from the Department of Technical Sciences of CAS. At the level of each province or municipality, and in keeping with the decentralized economy discussed above, there appears to be a science and technology bureau that is responsible for the research done within its jurisdiction and for seeing that scientific activities are coordinated with local economic needs.[78] Many of the former CAS institutes may now be under the authority of these bureaus. This would mean that, in an operational sense, each province has substantial policy-making authority and that China may now have a series of provincial science policies as well as a national policy framework provided by the center. This situation differs from the 1957-1961 period in that the provincial science bureaus appear to have far more authority now in the absence of the powerful centralizing forces provided by the SSTC.

Within the Academy itself, the old guard of intellectual leaders under President Kuo Mo-jo, has been thinned more by old age and natural death than by politics. The chief officer for academic matters appears to be aging physicist Wu Yu-hsun, a long-time Vice-President of CAS. The Party and administrative leadership exercised through the Secretariat, however, has changed. Party Secretary Chang Ching-fu has apparently been permanently purged. The top politically responsible leader in the Academy at present appears to be Yueh Chih-chien, another former Vice-Chairman of SSTC, and former director of the Cadres Bureau. Yueh seems to be performing the functions of the former role of Secretary General.[79]

It appears that the old system of having separate committees for administration and academic affairs has given way to a single revolutionary committee embodying the "3-in-1 combination" of revolutionary cadres, workers and technicians, and progressive intellectuals. In many cases the cadres and intellectuals are the same people who held leadership positions before the Cultural Revolution. Indeed, by 1972, reports indicated that the Party organizations in most research units were once again the ultimate source of authority. The new format, however, in principle provides for greater participation in decision-making for workers, technicians, and younger scientists.

The relationships among CAS, the science and education group, and the provincial science bureaus are difficult to discern at present due to the continuation of uncertain center-province political relations and to a lack of specific information. It could be that the Science and Education Group will emerge as a powerful central body comparable to the old SSTC if recentralization trends continue. A more likely outcome, however, would be that the Science and Education Group and CAS will concern themselves more with advanced science and education, and with more sophisticated technology serving the needs of centrally controlled industry and defense. The provincial science establishments, on the other hand, will become more involved with research in the service of intermediate and adaptive technologies and will employ organizational arrangements conducive to realizing the Maoist ideal of integrating research, education, technical innovation, and production.

The long-standing concern for making research relevant to production received some interesting new organizational expressions since the Cultural Revolution. The intense reeducation of intellectuals, the encouragement of worker and peasant participation in research, and the 3-in-1 combination are all intended to create this desired linkage by breaking down organizational barriers between research and production. One new approach practiced by some research units is the "3-3 system." Under this system, one-third of the units' personnel go out into production settings to make studies and provide advice. Another third stays in the laboratories to do research. The final third spends longer periods in production settings, engaging in labor, and making their talents available on a day-to-day basis. The three groups rotate these assignments. The experiences of groups 1 and 3 are fed back into research activities, hopefully insuring the relevance of the latter.[80]

Universities have been urged to establish linkages with relevant production units. Chiaotung University in Sian, for instance, practices the principles of "going out" and "inviting in." "Going out" involves sending teams to factories to conduct research and work with the workers and factory technical staff. "Inviting in" involves providing the technical facilities of the university to groups of workers and technicians from the factories.[81] Interestingly, however, this kind of collaboration often involves organizations from different administrative jurisdictions. The program, therefore, requires coordination from a higher level, which may result in a gradual recentralization of some important areas of decision making.[82]

The model that has evolved since the Cultural Revolution began can be referred to as "mobilization model II." Table 4-3 compares the five models discussed in this chapter.

The alternative models for the organization and administration have been closely linked with the major shifts in national policies discussed at the outset. It will be useful to compare these shifts in science policy and implementation against the theoretical considerations outlined in Chapter 1. Before doing so, however, we must look at another dimension of China's scientific development—her efforts to mobilize the Chinese masses for participation in experimental and innovative activities.

Table 4-3
Competing Models for the Organization and Administration of Chinese Science Programs

Organizational and Administrative Factors	Professional Model	Bureaucratic Model
1. Advocates	Scientists.	Party and administrative cadres.
2. Leadership and Party Influence	Leadership belongs to scientists of proven ability; grudging acceptance of administrative leadership in providing the material conditions for research, Party influence generally opposed except to the extent that it can facilitate the supply of the material conditions for research.	Primacy of leadership by non-scientists; separation of administrative and Party leadership in principle, but with recognition of the ultimate authority and primacy of Party leadership.
3. Centralization	Decentralized to level of experienced individual scientist.	Policy and administrative leadership centralized in SPC; academic leadership centralized in CAS.
4. Differentiation	Favored; but on basis of skill; (e.g., applied researchers should do applied research, basic researchers should do basic research regardless of place of employment).	Favored; but on the basis of the specialization of functions within a network of organizations (e.g., CAS as locus of basic research).

Mobilization Model	Bureaucratic-Professional Model	Mobilization Model II
"Maoist" or "radical" elements in the CCP.	Technocratic coalition of non-Maoist Party and government cadres, and influential scientists.	Maoist Party and government cadres, younger scientists, technicians and workers.
Principle of Party leadership literally and rigidly enforced; Party leadership extends into academic matters and influence of scientists in decision-making sharply curtailed.	Principle of Party leadership upheld but political interference with science greatly reduced as Party becomes bureaucratized; science seen as complex activity to be administered (not politically mobilized) by specialists who bridge professional and organizational interests; at institute level, separate leadership committees for administration and academic affairs.	Principle of Party leadership de-emphasized as a result of attacks on bureaucratized Party during Cultural Revolution; leadership by revolutionary committee embodying "3-in-1 combination," the meaning of which varies, but usually includes representation from (1) the masses, (2) revolutionary cadres, (3) progressive scientists and technicians; dual committees system at institute level replaced by single committee reflecting the 3-in-1 principle and the blurring of distinctions between politico-administrative and technical questions.
Simultaneous centralization and decentralization. Policy-making authority, and areas of more sophisticated research highly centralized under SSTC and CAS and central ministries respectively; nevertheless, growth of a regionalized research network including provincial branches of CAS and "research groups" attached to production levels regionalized network under administrative supervision of provincial Party Committees.	Highly centralized policy-making planning and administration under organs such as SSTC and Cadres Bureau; regional expansion of research network undertaken cautiously (1957-1961 expansions consolidated).	Decentralization to provincial authority of many units formerly controlled by the Center. A centrally controlled sector of the science system remains under CAS and "Science and Education Group."
Under attack; scientists are to become workers, workers are to become scientists; organizational boundaries attacked as impediments to release of mass initiative; factories and farms become	Differentiation seen as part of administrative rationality; based both on skills and capability, and functional specialization within the science system.	Differentiation strongly opposed as in first mobilization model; differentiation linked to stratification society and destratification takes priority over even the most rationally justified differentiation.

Table 4-3 (cont.)

Organizational and Administrative Factors	Professional Model	Bureaucratic Model
Item 4. Differentiation *(cont.)*		
5. Recruitment	Controlled by scientists and based on demonstrated competence and ability.	Controlled by administrators/ Party and based upon organizational needs and procedures.
6. Personnel Mobility	Free and subject only to market considerations and individual interests and aspirations.	Administratively controlled and based upon organizational needs and procedures.
7. Rewards & Incentives	Although adequate compensation is expected, most meaningful rewards come from colleague recognition from the international scientific community; incentives therefore to achieve that recognition through publication; some desire to serve the nation also present but subordinate to concern for recognition.	High salaries provided indicating high value regime places on scientists; state sponsored awards for achievement; efforts to undermine professionally relevant rewards.
8. Project Selection	Function of the internal development of the science; influence of the "paradigm"	Function of the needs of the state as outlined in the plan.

Mobilization Model	Bureaucratic-Professional Model	Mobilization Model II
not only units of production, but units of education and research as well; nevertheless, some differentiation maintained as the "two legs" strategy implies and as establishment of new institutions like University of Science and Technology indicates.		
Controlled by Party and based upon political and ideological considerations.	Controlled by administrators/Party in consultation with scientists, and based on criteria of intellectual achievement.	Controlled by "revolutionary committees" in close consultation with workers and peasants in production units; based on politics and ideology as well as intellectual promise.
Controlled by Party and based upon political and ideological considerations.	Administratively regulated by center on basis, of individual capability and interests, and needs of the science system.	Controlled by Party at the province level, or by Party center for the centralized sector; based on political and ideological considerations as well as local economic needs.
Although there is no indication that high salaries for scientists have been abolished they are greatly de-emphasized; the reward system of the scientific community (i.e., colleague recognition) is intensely criticized; state sponsored awards for achievement suspect; non-material, ideological incentives emphasized.	High salaries and state-sponsored prizes and awards; based on both organizational and professional criteria.	High salaries probably continue but special awards and prizes are criticized; sharp criticism of purely professional rewards; stress placed on ideologically defined rewards stemming from service of the people.[a]
Principle of science serving production emphasized; as organizational barriers break down and scientists are sent into productive milieu, they are, with cooperation of the masses, to choose topics and apply their knowledge and abilities in response to production needs; plan loses significance in that it is constantly revised and in this sense work becomes less routinized.	Function of planning expressing needs of the state; great reliance placed on scientists in formulating plans however; inputs from scientific community institutionalized in regular meetings of scientific societies and in specially convened national and regional meetings.	Operation of national planning uncertain and may be limited to very general policy guidance; project selection probably a function of local production needs as defined by 3-in-1 groups.

Table 4-3 (cont.)

Organizational and Administrative Factors	Professional Model	Bureaucratic Model
9. Project Control	Flexibility exercised by competent professionals.	Exercised by administrative cadre in consultation with professional.
10. Project Evaluation	Performed by leading scientists, and ultimately by larger scientific community.	Performed administratively in reference to plan.
11. Communication	Informal, through journals and professional conferences, opposed to secrecy.	Formalized according organizational procedures; secrecy justified in the interest of organizational integrity and national security; specialized mechanisms for information transfer and control established.
12. Foreign Collaboration	Support for principle of non-political internationalism in science; advanced training in capitalist countries experienced by many results in *de facto* orientation to literature from the West.	Internationalism in science linked to politics; "lean to one side" policy results in orientation to Soviet and Eastern European science; active attempts to discredit Western science.
13. Generational Relations	Younger men must respect and learn from older men.	Older men must pay particular attention to training younger men; youth should learn science from older men, but efforts must be made to prevent older men from becoming role models for youth.
14. Extra-Academic Meetings	Political meetings opposed as useless, grudging and reluctant participation in administrative meetings.	Political meetings regarded as necessary, but limited to one day/week; great attention to administrative meetings and many experienced scientists coopted as administrators.

[a]There are some indications that high salaries may still be in effect for older scientists but that a much lower ceiling on the salaries of younger men has been established.

Mobilization Model	Bureaucratic-Professional Model	Mobilization Model II
Exercised by Party; although lip service is paid to relevance of plan, ideological criteria predominate.	Exercised by professionals subject to review by administrative cadres.	Exercised by revolutionary committee.
Performed by Party in reference to production needs and ideological criteria.	Performed by combining professional and administrative criteria within framework of plan.	Performed by revolutionary committees employing their varying criteria within context of local production needs.
Attacks on stratification, differentiation and organizational boundaries intended to open up communications within the system; however the mandating of communication may result in psychological barriers to communication; secrecy still continues.	Professional modes of communication through journals and professional conferences given official support and thereby becomes more formalistic; the differentiated nature of the system makes satisfactory communication both difficult and essential.	"3-in-1 combinations" can be seen as institutionalized communications mechanism to insure proper flows of messages among production, policy, and research and education units; both the promise and the problems of the first mobilization model still obtain.
Greatly de-emphasized as native methods and native wisdom are celebrated; "cultural nationalism" to the fore.	Internationalism in science still linked to politics, but the Sino-Soviet rift results in more flexible interpretation of this link; interest in, and cooperation with non-Communist countries begins resulting in as overenthusiastic search for foreign knowledge and methods that violates principal of cultural nationalism.	Native methods and wisdom highly celebrated; traces of anti-foreignism give way to guarded enthusiasm for international cooperation and importation of foreign technology; cultural nationalism and self-reliance remain the cornerstone of policy however.
Generational differences exacerbated; youthful frustrations channeled into criticisms of elders as youthful rebelliousness is legitimized; distinction between older men as sources of scientific knowledge and as role model blurred.	Value of experienced scientists highly prized and their contributions to research and training of next generation emphasized; structure of system ensures their superior position over younger men.	Experienced scientists still prized, but are still in need of reform; their alleged conservatism and orientation to foreign science and technology makes them legitimate targets to be "struggled against" by younger generation; structure of system more egalitarian.
Indiscriminate political and administrative meetings disrupt research routines and cut into time available for research and teaching.	Required but held to a minimum; scientists' time rationalized by administrative means.	Research itself seen as political activity and political activity seen as part of science; new mode of experience being tested; much time consumed in reconstructing system along Maoist lines.

Notes

1. The term "model" is used here to refer to a conceptual-behavioral amalgam. It refers both to the way the various Chinese actors, scientists, administrators and policy makers have thought about science and its relationship to society and the actions that have resulted from the interaction of those thoughts with the exigencies stemming from political and economic realities.

2. CAS President Kuo Mo-jo stated that in 1952 there were 1290 researchers and assistant researchers in CAS. *New China News Agency (NCNA)*, September 17, 1958, in U.S. Consulate General, Hong Kong, *Survey of the China Mainland Press (SCMP)* 1737. Chu-yüan Cheng states that there were 164,000 engineers in 1952. See his *Scientific and Engineering Manpower in Communist China, 1949-1963* (Washington: U.S. National Science Foundation, 1965), p. 11.

3. These tasks give further insight into the targets of the twelve-year plan. They include: comprehensive expeditions (resource surveys), surveying and map making, oceanography and meteorology, geological mineral products, metallurgy, petroleum and coal, power, machine building, inorganic chemical engineering, heavy chemical engineering, light industry, building materials, hydraulic engineering, radio and communications and transport, atomic energy, aviation, calculating technology, electricity and supersonic technology, defense problems, agriculture, forestry, medicine, measurement, basic research (*chi-pen li-lun ho chi-ch'u hsueh-k'o*), and scientific information. Kuo Mo-jo, "On Coordination of Scientific Research," *Jen-min Jih-pao (JMJP)*, March 20, 1957, in *SCMP* 1515.

4. Ibid.

5. Ibid.

6. Ibid.

7. "Numerous Contradictions Within the Ranks of The Scientific Circle," *NCNA*, April 30, 1957, in *SCMP* 1541. The microbiologist mentioned, Fang Hsin-fang, was a member of the academic committee of the Institute of Applied Mycology, but he was not a member of the department committee of the Department of Biology. It is difficult to determine whether a department committee member would have suffered the same plight, but it may be that the department committee member, with far more influence in the planning process was capable of insuring that his own work would be included in the plan, hence, financially supported. It is not impossible that planning and the department system, with its elite membership worked to create a "scientific establishment" that produced a cleavage in the ranks of Chinese scientists. The recurring injunction to practice "academic democracy" may be a response to the development of social stratification within the scientific community itself.

8. Ibid.

9. "Criticisms and Suggestions Made by Scientists at Department Meetings of Department Committee of the Academy of Sciences," *JMJP*, May 27, 1957, in U.S. Consulate General, Hong Kong, *Current Background (CB)* 460.

10. "Academy of Sciences Convenes Forum for Peking Scientists to Discuss Contradictions Within Their Ranks," *NCNA*, May 3, 1957, in *SCMP* 1541. The

criticism of the awards was made by Yuan Han-ch'ing (Ph.D. University of Illinois), a scientist of some standing, who referred to the rewards as "the purchase of passiveness with cash." There is the possibility that Yuan, who was not an award winner, might be expressing feelings of sour grapes. A more likely explanation, however, is related to the ambiguity of the awards themselves. If the awards were a reflection of competent and genuine colleague recognition of achievement, it is likely that scientists would support them even if they were not winners, because of their commitment to the norms of the professional community. Considerably less professional enthusiasm would be expected for awards that are offered as incentives by the state.

11. "Numerous Contradictions Within the Ranks of the Scientific Circle." The issue of bureaucratism received much attention at this time and was a charge that was apparently leveled at some scientists as well as administrative cadres. In words that could have been spoken in Washington or Moscow, Kuo Mo-jo, CAS President, defended these scientists and, as a necessary annoyance, the support bureaucracy of modern research as well. Kuo noted:

> Here I wish to salute . . . those of our comrades engaged in administrative and office work whose organizational ability is comparatively strong. Their work is valuable but their tasks are thankless. If they do their jobs well, they are merely doing their duty; but if they make the slightest mistake, everyone will jump on them.
>
> It seems that we should give them our thanks and respect, particularly those who are scientists themselves, but who because of the needs of the state are compelled to take up administrative or office work in service of scientific research. . . . These respectable friends, so far as I know, do not love to be government officials. They too want to return to the ranks of scientists. The term 'bureaucracy' exposes its members to odium, but since you have a state and a government, you cannot help but take part in its management. Science administration is an important part of the business of the state, and it will not do to leave it to chance . . .

"Closing Address" (given at Second Plenum of Department of Committee), *JMJP*, May 31, 1957, in *CB* 460.

12. Statement by metallurgist Kung Tsu-t'ung in "Criticisms and Suggestions made by Scientists at Department Meetings of Department Committee of the Academy of Sciences."

13. The organizational and administrative factors in the table are understood in the following terms:

1. Advocates–those actors in the science system who serve as spokesmen for a particular model.
2. Leadership and Party Influence–actors making key decisions on policy, administrative and technical matters.
3. Centralization–the locus of decision making on policy, administrative and technical issues.
4. Differentiation–the extent to which roles and organizations within the science system are distinguished on the basis of specialized functions.
5. Recruitment–the process of selecting new scientific and technical personnel,

the criteria employed, and the control over the process.

6. Personnel Mobility—the process of moving scientific and technical personnel from one job to another, the criteria employed, and the control over the process.
7. Rewards and Incentives—devices used to motivate scientific and technical personnel to achieve conformity with certain kinds of incentives.
8. Project Selection—the criteria employed in choosing one research or development project over another.
9. Project Control—the mechanism used for overseeing the conduct of a research or development effort.
10. Project Evaluation—the process of assessing the worth of a project, the criteria employed, the control over the process.
11. Communication—the flows of information within and between organizations within the system, between the system and user of science and technology, and the factors affecting those flows.
12. Foreign Collaboration—the relationships of the Chinese science system with science and technology in the international environment.
13. Generational Relations—the nature and functions of the relationships between older and younger scientific and technical personnel.
14. Extra-Academic Meetings—administrative and political meetings which consume time which might otherwise be invested in research and development activities.

14. This was particularly true in regard to expanding research beyond CAS to production ministries, and especially universities, and in providing material resources and support services for research. The scientists' desire for substantially more autonomy, however, were not met. "Conclusion of Enlarged Meeting of the Scientific Planning Commission of the State Council," *NCNA*, June 15, 1957, in *SCMP* 1564.

15. The role of nationalism in Chinese thinking about technological development has been nicely discussed in the work of Rensselaer W. Lee III. See his "Ideology and Technical Innovation in Chinese Industry, 1949-1971," *Asian Survey* XII 8 (August 1972), pp. 647-661; and "The Politics of Technology in Communist China," in Chalmers Johnson (ed.), *Ideology and Politics in Contemporary China* (Seattle: University of Washington Press, 1973), pp. 301-325.

16. See Franz Schurmann, *Ideology and Organization in Communist China* (Berkeley: University of California Press, 1970), pp. 193-205.

17. *NCNA*, November 28, 1957, in *SCMP* 1670.

18. *NCNA*, February 20, 1958, in *SCMP* 1728.

19. *NCNA*, March 12, 1958, in *SCMP* 1735.

20. Chang Ching-fu, "For the Liberation of Thought and the Determined Follow Through of the Main Party Line Concerning the Construction of Socialism in the Tasks of Science," *JMJP*, June 7, 1958, in U.S. Department of Commerce, *Joint Publications Research Service (JPRS)* DC 472.

21. Fan Ch'ang-chiang, "The Question of 'Motive Power' of the Intellectuals," *Chung-kuo Ching-nien*, June 1, 1958, in U.S. Consulate General, Hong Kong, *Extracts from China Mainland Magazines. (ECMM)* 142.

22. Nieh Jung-chen, "The Realization of a Big Leap Forward in Science," *NCNA*, March 14, 1958, in *SCMP* 1745.

23. See Richard P. Suttmeier, "Chinese Scientific Societies and Chinese Scientific Development," *The Developing Economies* XI 2 (Autumn 1973).

24. Cheng, *Scientific and Engineering Manpower in Communist China*, pp. 21 and 81.

25. Perhaps the best statement of the "walking on two legs" strategy as it applies to science policy is Nieh Jung-chen. "Let the Whole Party Lay a Firm Hold on Scientific and Technological Work to Carry Out the Technological Revolution," *JMJP*, August 2, 1958, in *SCMP* 1831.

26. Yu Kuang-yuan, "On Mass Movements in Scientific and Technological Work," *Hung Ch'i* 3 (February 1, 1960), in *JPRS* 3169.

27. Ibid.

28. Li Tsai-ming, "How Forward Leaps were Accomplished at the Laboratory of Semiconductors," *K'o-hsueh T'ung-pao (KHTP)* 18 (September 27, 1958), in *JPRS* 1479.

29. One case against the older scientists is made by SSTC Chairman, Nieh Jung-chen. According to Nieh, "it has been found that some scientists are conservative and backward at their jobs. . . . As soon as they have achieved a certain social position, many people would not seek further advancement. They are devitalized and cling tenaciously to their old and out of date knowledge. They no longer engage themselves actively in research work, and moreover would not even read new publications. There are even those who, though they are already lagging behind in learning, are not aware of the fact and do not try to catch up with others in front, but on the contrary . . . reject correct suggestions, suppress the growth of youthful forces, and insist that the younger scientists should work according to their old out of date methods, thus, obstructing the formal progress of science. . . . Some young people have achieved good results in their research work, but unfortunately the results were achieved not under the active support and assistance of the scientists concerned in the supervision of the research work, *but by struggling against them* or by purposely concealing the work from them. . . . *We must resolutely support our young scientists in their struggle against conservatism*" [italics added]. Nieh Jung-chen, "The Realization of a Big Leap Forward in Science." Cheng Chu-yüan has noted that promotions and opportunities for upward mobility generally have been denied younger scientists. Thus there may be other than ideological reasons motivating the younger men to struggle against their elders. See *Scientific & Engineering Manpower in Communist China*, p. 175.

30. For a negative evaluation of the mobilization model in operation, see Mikhail A. Klochko, *Soviet Scientist in Red China* (trans. Andrew MacAndrew), (New York: Fredrich A. Praeger, 1963).

31. Kang Chao, "Policies and Performance in Industry," in Alexander

Eckstein, Walter Galenson, and Ta-chung Liu (eds.), *Economic Trends in Communist China* (Chicago: Aldine Publishing Company, 1968), p. 564.

32. Changes in the use of central financial and commercial controls are nicely summarized in Franz Schurmann, "Politics and Economics in Russia and China," in Donald W. Treadgold (ed.), *Soviet and Chinese Communism: Similarities and Differences* (Seattle: University of Washington Press, 1967), pp. 297-326. See also Dwight H. Perkins, "Industrial Planning and Management," in Eckstein, Galenson, and Liu, op. cit., pp. 597-636. Perkins notes that the decentralization of the "Great Leap" was "subject to no constraint other than those of the Party discipline and ideology" (p. 628).

33. See Stephen Andors, "Revolution and Modernization: Man and Machine in Industrializing Society, the Chinese Case," in Edward Friedman and Mark Selden, *America's Asia* (New York: Vintage Books, 1971), pp. 393-444.

34. Charles Hoffmann, *Work Incentive Practices and Policies in the People's Republic of China, 1953-1965* (Albany: State University of New York Press 1967), p. 80.

35. See Carl Riskin, "Small Industry and the Chinese Model of Development," *The China Quarterly* 46 (April-June 1971), p. 261ff.

36. Anthony M. Tang, "Policy and Performance in Agriculture," in Eckstein, Galenson, and Liu, op. cit., p. 496.

37. G. William Skinner, "Marketing and Social Structure in Rural China, Part III," *The Journal of Asian Studies* XXIV 3 (May 1965), pp. 363-399.

38. See Tang, op. cit.

39. See Schurmann, "Politics and Economics in Russia and China," and Chao, op. cit., pp. 566-7. On various meanings of centralization and decentralization and their implications for management see Schurmann, *Ideology and Organization in Communist China*, pp. 195-210.

40. See C.S. Chen (ed.), *Rural People's Communes in Lien-Chiang* (trans. Charles Price Ridley), (Stanford: Hoover Institution Press, 1969), and generally, Skinner, op. cit.

41. Thomas G. Rawski, "Recent Trends in the Chinese Economy," *The China Quarterly* 53 (January-March 1973), p. 24.

42. Audrey Donnithorne, *China's Economic System* (New York: Frederick A. Praeger, Publishers, 1967), pp. 102 and 490.

43. Jan S. Prybyla, *The Political Economy of Communist China* (Scranton, Pa.: International Textbook Company, 1970), p. 366.

44. See "A Criticism of the Theory of Making the Electronics Industry the Center," *JMJP*, August 12, 1971; "Deliberate Castration and Shameless Betrayal–A Criticism of Liu Shao-ch'i's and Other Political Swindler's Theory of 'Electronics as the Core,' " *KMJP*, December 13, 1971, in *SCMP* 5045.

45. *Kung-jen Jih-pao*, February 23, 1962, in *SCMP* 2699.

46. For a good discussion of industrial development strategies, see "Industrial Development of China," *Current Scene VI* No. 22 (December 20, 1968).

47. Leslie T.C. Kuo, *The Technical Transformation of Agriculture in Communist China* (New York: Praeger Publishers, 1972), pp. 22-23.

48. See *KMJP*, August 20, 1961, in *SCMP* 2590; *JMJP*, September 5, 1961, in *SCMP* 2580; and *KMJP*, September 20, 1961, in *SCMP* 2598.

49. For a discussion of this type of leadership, see Lo Ying, "Achieve Rejuvenation by Our Own Efforts, Make Haste to Catch Up with the Advanced," *KMJP*, April 30, 1963, in *SCMP* 2983.

50. For a statement of guidelines, see "Actively Unfold Scientific Study and Discussion of Technological Policies," *Hung Ch'i* 11 (June 1, 1962), in U.S. Consulate General, Hong Kong, *Selections From China Mainland Magazines (SCMM)* 319.

51. See Ho Tsu-hsiu, "The Roles of Experiments, Abstractions, and Hypotheses in Scientific Research," *Hung Ch'i* 11 (June 1, 1961), in *SCMM* 266; Ho Tsu-hsiu, "Natural Science and Practical Application," *Hung Ch'i* 7 (April 1, 1962), in *SCMM* 311; *KMJP*, December 22, 1962, in *JPRS* 18023; Wu Chung-wei, "The Procedure Problem in Scientific Research Work," *KMJP*, June 4, 1964, in *JPRS* 25, 337; Ch'ien Hsueh-sen, "Large-Scale Scientific Experimental Work," *JMJP*, August 30, 1964, in *SCMP* 3297; Kung Yu-chih, "On Scientific Experiment," *Hung Ch'i* 1 (January, 1965), in *SCMM* 454; and Mao I-sheng, "Knowledge and Experience of Scientific Experiment," *Hsin Chien-she* 11-12 (December 20, 1965).

52. See particularly "Correct Way of Conducting Agricultural Scientific Experiments," *JMJP*, July 23, 1963; and Yu Ch'i-pao, "Some Points Concerning Research to be Carried Out Jointly by Agronomists and the Masses," *Hung Ch'i* 22 (November 19, 1963), in *JPRS* 22660.

53. For a Soviet justification of science as direct production force, see I.G. Kurakov (trans. Carin Dedijer), in *Science, Technology and Communism, Some Questions of Development* (Oxford: Pergamon Press, 1966).

54. *Chung-kuo Hsin-wen*, July 15, 1964, in *JPRS* 26289; and *JMJP*, December 2, 1963, in *SCMP* 3117.

55. The apparent fascination with the research-intensive electronics industry mentioned above may be indicative of the role of science in China's industrial development envisioned by Liu and his supporters. See note 46.

56. Ch'ien Hsueh-sen, "Organization and Management Work in Science and Technology," *Hung Ch'i* 22 (November 19, 1963), in *JPRS* 22660.

57. William Kornhauser, *Scientists in Industry's Conflict and Accommodation* (Berkeley: University of California Press, 1962).

58. See "Regulations on Awards for Inventions," *JMJP*, December 2, 1963, in *SCMP* 3117; and "Encourage Inventions and Technical Improvements by Rewards in Order to Promote Development of our Production and Construction," *JMJP* (editorial), December 2, 1963, in *SCMP* 3117.

59. For a fuller discussion of this point, see Suttmeier, "Chinese Scientific Societies and Chinese Scientific Development."

60. "Raise the Level of the Academic Activities of Natural Science Societies," *KMJP* (editorial), September 9, 1962, in *SCMP* 2832.

61. Ibid.

62. "A Year of New Achievements and Many Successes in Research for Sci-

entists in China," *NCNA*, January 13, 1963, in *SCMP* 2906.

63. "Academic Activities in China," *KHTP* 3 (March 17, 1962), in *SCMM* 313.

64. See particularly, "People Use Mao Tse-tung Thought to Study Solar Eclipse," *NCNA*, January 24, 1969, in *SCMP* 4349; also "China's Success in Making Large Scale Comprehensive Observation of a Total Eclipse" and "Background Material: Solar Eclipse and Their Research," *NCNA*, January 23, 1969, in *SCMP* 4348.

65. See Cheng, *Scientific and Engineering Manpower in Communist China, 1949-1963*, pp. 277-279.

66. This meeting was reported a year later in *K'o-chi Chan-pao (Science and Technology Combat News)*, June 2, 1967, in *SCMP* 4011.

67. Ibid.

68. Ibid.

69. This is not to say that certain administratively prominent scientists were not attacked, nor that the persistence of "bourgeois" *ideology* among the scientists was ignored.

70. *NCNA*, October 16, 1966, in *SCMP* 3805.

71. *NCNA*, April 13, 1967, in *SCMP* 3920.

72. Shih Yen-hung, "Down with the Chief Backer of the Revisionist Educational Line," *JMJP*, July 18, 1967, in *CB* 836.

73. See Bernardo Kordon, "Kuo Mo-jo Discusses the Cultural Revolution," *Clarin* (Santiago), February 9, 1969, in *JPRS* 47685.

74. For an interesting account of the experiences of scientists in the May 7 schools, see *JMJP*, June 24, 1972, in *SCMP* 5168.

75. *NCNA*, July 22, 1968, in *SCMP* 4227.

76. For a detailed account of factory management in one plant, see Mitch Meisner, "The Shenyang Transformer Factory—A Profile," *The China Quarterly* 52 (October-December, 1972), pp. 717-736.

77. See, for instance, "Mao Tse-tung Illuminates the Road of Achieving Greater, Faster, Better and More Economical Results in Geological Work," *JMJP*, February 14, 1971, in *SCMP* 4846. See also Audrey Donnithorne, "China's Cellular Economy: Some Economic Trends Since the Cultural Revolution," *The China Quarterly* 52 (October-December, 1972), pp. 605-619.

78. Private communications to the author from recent visitors to China.

79. *Current Scene* IX No. 8 (August 7, 1971), p. 17.

80. See *KMJP*, May 18, 1972, in *SCMP* 5143; and *NCNA*, June 26, 1972, in *SCMP* 5170.

81. *JMJP*, October 21, 1971.

82. See, P'ei Yu, "Establish A Good Link-Up Between Factory and School in an Earnest Way," *Hung Ch'i* 9 (September 1, 1972), in *SCMM* 737-38. P'ei Yu noted that these link-ups may need central support and states that the center should not shirk its responsibilities. Interestingly, P'ei is on the staff of the central 4th Ministry of Machine Building.

5 Technological Mobilization: Science, Technology, and the Chinese Masses

Fortune favors prepared minds.

—Louis Pasteur

As we have seen, the Chinese Communist elite, at the time of the founding of the People's Republic, displayed a great amount of enthusiasm for science and technology. The Common Program of 1949 stressed the importance of science and technology not only for creating an economically and militarily strong state, but also for transforming the cultural basis of Chinese society into one in which the methods and spirit of science would be pervasive. While effort and resources were being invested in research and development to meet the first objective of a strong China, efforts were also being made to meet the second.

The importance of this second objective was clearly evident in Chinese ideology. In the emerging new socialist state, workers were to be the masters of the technology used in production. However, an active research program was expected to produce changes in technology. For workers to continue to master the technology in their work environment, it would be necessary to continue to raise the level of their technical knowledge and skills.[1] A second ideological factor, which influenced Chinese thinking about the relationship of science and technology to the masses, was the close association that ideology asserted between science and production. According to this view, science had its origins and its ultimate verification in production, and the masses of workers and peasants were most intimately familiar with production.

From the point of view of the non-Chinese, non-Marxist outsider, the attention given to instilling in Chinese society a positive orientation toward science and technology can be viewed as one technical development strategy for a problem common to many low-income countries. The problem is one of making science and technology relevant to developmental needs. Although it now seems obvious that the mere establishment of research and development organizations will not insure the application of science and technology to production, it was not always so.

Far too often research facilities have been established with little thought to the social environment in which research was to be conducted and the mechanisms necessary for linking research with environment. However, the societal characteristics of most low-income countries, which comprise the environment of such research establishments, have usually included neither an appreciation nor a demand for the outputs of research. In part this has been due to the irrelevance of those outputs for developmental needs, but it has also been due to the failure of those closest to production to identify and take advantage of technological opportunities. Such technological opportunism is associated with a spirit of entrepreneurship, a spirit which is unfortunately not always present in great abundance in low-income countries.

China has not found a magic formula for making science and technology relevant for her needs. She has, however, given serious attention to making the economic milieu innovation-oriented and to mechanisms for linking research with production. She has devised various programs at various times, some of which have endured, some of which have not. The fact that debates over, and injunctions about, how science should serve agriculture or how innovation programs should be organized in factories have surfaced 15 to 20 years after these problems were initially attacked indicates that optimum solutions have not yet been found. This chapter is an analysis of the evolution of some of these programs.

On the Concept of "Technological Mobilization"

Analytically, the problem of bringing science and technology to bear on the problems of production involves three processes. First, usable information must be supplied to a potential user. The information may come in a variety of forms from a variety of sources. For our purposes, a new machine or a blueprint may be understood as information, as is a scientific theory, or an agricultural technique that has been locally perfected by peasants over time. Second, resources must be available to procure and/or utilize the information. Most importantly, however, information and resources must be manipulated in such a way that innovation occurs—that is, that information is applied in a new way in actual production. This action product of the combination of information and resources is a complex phenomenon which we may call "technological mobilization."

Technological mobilization can be thought of as a variant of Karl Deutsch's concept of "social mobilization":

> . . . the process in which major clusters of old social, economic, and psychological commitments are eroded or broken and people become available for new patterns of socialization and behavior.[2]

Technological mobilization then entails breaking commitments to old artifacts, processes, and patterns of organization, making people available to be introduced to new technologies, and making them receptive and eager to try and adopt these. Technological mobilization is an educational process in which communication is of the greatest importance, but it must also be inspirational and reinforcing as well, since, as many students of innovation have noted, those being mobilized perceive risk.[3]

Again, analytically, the barriers to innovation can be readily identified. Anything that impedes the flow of information is clearly a barrier. This could range from a simple absence of needed information in a system to poor communication within a system. The poor communication may be a function of spatial obstacles, organizational obstacles, or problems of psychological distance between those with the information and those who would use it. Similarly, the non-availability of resources is a barrier. Resources can take a variety of forms that may include money for salaries or for experimental materials or non-monetary materials for experiments and demonstrations. In any social system, a certain amount of slack in available resources must be identified for diversion to innovation.

A final type of barrier to innovation concerns the creative interaction of information and resources. Barriers here include the absence of incentives to innovate or the presence of disincentives. Another barrier can be the organization of authority in the innovation situation. Excessive hierarchy can result in speedy but superficial innovation, while a colleagial or collective pattern of authority may insure more lasting and complete technological mobilization.[4]

Barriers to innovation in socialist planned economies are a special case of the above. Berliner, on the basis of Soviet experience, has argued that there exists a basic conflict between innovation and the performance criteria used to evaluate plan fulfillment. Soviet managers perceive the rewards for plan fulfillment as more attractive than bonuses for innovation. Berliner offers the following reasons for this situation:

1. There is typically a time-lag before an innovation begins to pay off, during which output may be lower than before the introduction of the innovation.
2. The tenure of office of Soviet managers is relatively short. They may therefore not be around to enjoy the benefits of the innovation, nor to be recognized for it. They may instead become recognized as managers who under-fulfill plans.
3. Innovations typically require new supplies and resources, which are perennially scarce in the Soviet economy.
4. Adjustments in the prices of new products often result in a lower rate of profit than for old products and "count for less in computing plan-fulfillment."
5. New equipment requires new spare parts, and consumers are often reluctant to install new equipment for fear of not having spare parts in the event of breakdown.
6. The introduction of an innovation often requires "extensive testing and negotiations with research institutes and other official agencies whose approval is required before the production is authorized."
7. Ill-advised campaigns for innovation in agriculture have reduced the confidence of agricultural officials in new equipment and techniques promoted from the industrial sector and bureaucracy.
8. The planning bureaucracy does not enforce innovation assignments, and generally reinforces the conservatism of the managers.[5]

Although *all* of these reasons may not be relevant to China, they do point to the problems of creating an innovative environment in a highly structured, plan-oriented system of economic management, which did exist in China as a result of economic decisions made by the Chinese in the early 1950s.

Technological Mobilization and Social Change in China

Questions relating to technological mobilization take us back to a number of concerns discussed at the outset of this study. In particular they pertain to the relationship between technology and social change. Western social theorists have often spoken of the "impact" of technology on society, and of technology inducing changes in society. The implication is that technology *causes* social change, as society (individuals and organizations) adjusts itself to technology. On the other hand, as our discussion in Chapter 1 has indicated, social change within given units of organization is often a prerequisite for technological advance. Although it is tempting for the Western mind to view these two technology-social change formulas as contradictory, they may not be. At this point we will take the easy way out and assume that both are true, and if causation must be specified, a pattern of circular causation may be assumed.

From the point of view of the Chinese, we might expect a fairly straightforward explanation of the technology-social change relationship in terms of changes in the productive forces as a primary cause giving rise to changes in the relations of production and superstructure. The Chinese, however, have not been able to resolve the question so simply. Indeed since the end of the period of socialist transformation in 1956, the issue of the relationship between technology and social change seems to have been a fundamental, underlying point of cleavage within the leadership. A discussion of this issue will show that technological mobilization, like the organization and administration of science, has been subject to major policy shifts. First, however, let us examine briefly some of the approaches to technological mobilization employed in the 1949-1957 period.

In the first period of Chinese technological mobilization, 1949-1957, the following strategies were employed: technical innovation campaigns, the identification and nurturing of technical activists, technical nucleation, extension activities, and the popularization of science. This first period is characterized by its experimentation and lack of overall coordination and integration. Although the programs were intended to begin the cultural transformation of the masses and to improve production, and probably achieved a certain success, their greatest significance may have been to demonstrate the regime's commitment to science and technology and the value it placed upon innovation.

Technical Innovation Campaigns[6]

Although these campaigns take on far greater significance during the "Great Leap Forward", Rensselaer Lee has reported that they have their origins in the immediate pre-liberation period and continue into the early post-liberation era.[7] Although unaccompanied by the ideological fanfare of the "Great Leap" period, the early campaigns urged workers to take initiative to improve old machines, tools, and processes, and/or to invent new ones. These campaigns were supported by trade union organizations that acted as "advocates of workers' innovation proposals before the factory administration."[8] It should be recalled that during this period, factory administration was based on the Soviet principle of "one-man management," under which skilled managers were given responsibility and were held accountable for the achievement of production targets. It is not surprising, therefore, that during the First Five-Year Plan, the momentum for worker innovations was lost as managers sought to meet the targets of the plan and as they became conscious of the inefficiencies of indigenous innovations in the face of imported technology from the Soviet Union.[9]

Technical Activists

Programs to identify, encourage, and nurture technical activists are closely related to worker innovation campaigns.[10] The identification and popularization of the names of model workers was one technique used in the innovation movement. Yet, the encouragement of technical activists deserves special mention because it represents China's attempt to generate a functional analog to the spirit of entrepreneurship which plays so important a role for technical advance in capitalist systems.[11]

The numerous accounts of technical activists in the Chinese press typically include the following characteristics: the technical activist (whether in the factory or on the farm) has a technical aptitude for inventions, rationalizations, designs, and experiments; he was frustrated in the old society because his aptitude was not recognized; his spirit has been released in the new society where he now spots new technical opportunities, takes the initiative to turn opportunities into technological progress, and shows perseverance when his ideas are criticized by established technical and administrative authority and by his peers who feel his aspirations exceeded his station. The technical activists often receive support from the Communist Party, which serves as their advocate in decision-making; through Party support more sophisticated technical advice is made available to them, and typically they receive further education and training leading to eventual promotion.[12]

The encouragement of technical activists is an appealing strategy for technological mobilization. The technical opportunity identified by the activist often seems unreachable to his peers who often ridicule his proposal. The activist's creativity and persistence, plus some scientific or technological insight from outside, combine in new ways to realize the technical opportunity. The realization of one opportunity, which is recognized as beneficial to all, demonstrates that formerly unattainable objectives can be reached by this combination of knowledge, creativity, initiative, and persistence. The technical activist is usually a worker or a peasant, which means that other workers and peasants can identify with him as one of their own, and not some specialized scholar whose competence lies beyond them. The fact that many technical activists have seen their lot in life improve—they are often promoted, held up as models, and since the Great Leap Forward, admitted to academic organizations—may induce others to become technical activists also.

Technical Nucleation

A third approach to technological mobilization can be called technical nucleation, the process of using a technologically advanced unit of production as a nucleus for demonstrating and aiding in the diffusion of technology to less developed units of production. State farms served this purpose in agriculture. They received the bulk of technical inputs into agriculture, and were expected in return to send out technical guidance teams to propagate improved seeds, pest control techniques, and other scientific farming techniques in the villages. In industry, technologically advanced plants were expected to help train workers from, and provide technical services for, other plants in the area as in the cases of the Anshan and Wuhan iron and steel complexes.[13]

Agricultural Extension

Apart from the use of state farms to aid in the diffusion of the technology to agriculture, China has also experimented with a number of programs that generally fall into the category of extension work. One aspect of this work was to extend the agricultural bureaucracy under the central Ministry of Agriculture down to subprovincial levels. The central Ministry for instance would direct seed selection and improvement efforts and then promote them through numerous seed-producing centers around the country that would provide seeds and advice on planting.

Another approach was to gradually develop a network of agricultural research organizations and experimental stations. By late 1955, this network consisted of 10 research institutes directly under the Ministry of Agriculture, 32 provincial experiment stations, 34 agricultural colleges, and a large number of subprovincial agricultural experiment stations.[14]

Of far greater long-term significance, however, were attempts to penetrate rural society at the level of the agricultural cooperative. Agricultural scientists and technicians were sent out from research institutes to spend time in the villages and conduct simple research on local production problems and demonstrate the rudiments of scientific method.

The accounts of these change agents present a familiar story to the student of technological diffusion. The scientist or technician from outside initially had difficulty getting his ideas understood and accepted. This was due in part to what sociologists have called "heterophily," the differences in values, beliefs, social status, education, and general life experience between interacting individuals.[15] It was also due to the fact that the science and technology carried by the change agent was not immediately applicable to the village situation. The invention or discovery to be transmitted may have been made under conditions that did not obtain in the village. In such a situation, the change agent would suddenly find himself in an insecure position, since the "trusted recipes" he had brought with him were not applicable, and he would be forced to live by his wits in an environment for which his socialization had not prepared him. He would have to become adaptive and creative with his knowledge, on the spot. It makes sense for him to have taken the Party's ideological injunctions seriously—he had to learn from the accumulated experiences of the peasants about local conditions and then "sum up" these experiences and his own knowledge.

The situation of the change agent in the rural milieu would be one of uncertainty. It was a situation where good leadership was important and where a local cadre could make the difference between success or failure. The Chinese rarely report the failures, although it is known that the quality of local leadership has been an abiding concern. In successful cases, however, the role of the local cadre is worth noting. He served as a bridge between the change agent and the peasants. In this role, he would have to induce the peasants to cooperate and be patient with the change agent and to value the information the change agent brought. This was important, since peasant suspicion of unproven innovations that might adversely affect their livelihood could result in non-compliance with the suggestions of the change agent. The cadre would also have to work with the change agent, encourage him, and induce him to learn from and respect the peasant. His job, in short, was to serve as a catalyst for the creative interaction of information and resources that is the essence of technological mobilization. An ability to bring about an environment where change agent, cadre, and peasant could participate in decision making facilitated such creative interaction.

A final aspect of these early attempts to send agricultural scientists and technicians to the villages was the establishment of local experiment stations in agricultural producers cooperatives (APCs) and later in production brigades. Where the change agents were successful, they left behind ongoing experimentation, which, however primitive, reflected a positive orientation to the possibilities offered by science and technology for improving production.

Popularization of Science

A final effort at technological mobilization during the initial period was the popularization of science. In 1950, the Chinese established the All-China Association for the Dissemination of Scientific and Technical Knowledge. The Association was to hold exhibits, develop children's literature on science and technology, assist in spare-time educational services, and develop films and popular science books and magazines.

During its first five years, the Association pursued popularization and dissemination activities which did not have the anticipated impact. On the occasion of its fifth anniversary, the Association was criticized for falling short of educating the people to help them increase production.[16] Accounts of worker and peasant innovation programs and of technical activists often noted how these activities were constrained by the technical level of the masses. These were natural situations for the work of the Association, but since there was no formal coordinating mechanism for bringing the resources of the Association to bear on these problems, opportunities were missed.

To remedy this situation, the Association began collaboration with the All-China Federation of Trade Unions (ACFTU) in 1956. The President of the ACFTU, Lai Jo-yü, noted that the demands of industrialization were pushing older, skilled workers into managerial positions. They were being replaced by less-experienced people with the result that the technical level of the work force was declining just at the time when a higher level of technological sophistication was needed.[17]

The weakness of the popularization program was symptomatic of Chinese efforts at technological mobilization during the period up to 1958. Although serious attention was given to bringing science to the masses, there apparently was no comprehensive and integrated effort to tie together the various approaches to technological mobilization. Just as the failure to coordinate worker innovation programs with popularization programs resulted in limiting the success of each, so the various extension efforts suffered by the absence of a rational organizational context that would have provided, on a regular basis, the necessary inputs for technological mobilization. Policy changes during the "Great Leap" began to remedy these problems. In spite of the problems, however, it does seem as if the regime established its commitment to science and technology, and to technical innovation, during this initial period.

It seems clear that there was a consensus within the Chinese leadership on the desirability of what we have called technological mobilization. However, a closer look at the various approaches to technological mobilization described above indicates two rather different premises underlying them. Nucleation, extension, and popularization are based on a "top down" philosophy of knowledge and information transfer from centers of expertise and more advanced technology to production settings characterized by low levels of expertise and technological sophistication. On the other hand, the technical innovation campaigns and the use of technical activists were founded on the premise that untapped expertise and creativity reside with those most intimate with the processes of production and that under proper leadership this potential could be realized. This might be thought of as the "bottom up" approach. Although it would seem to the outsider that both types of program would have some value, the differences in the underlying premises point to the more fundamental question of the relationship between technology and social change introduced above, a question which had implications for the characteristics of subsequent technological mobilization programs.

The movement toward the socialist transformation of industry and the increasing cooperativization of agriculture by 1956-57 seemed to indicate to the Chinese that they had been successful in fundamentally altering the relations of production in the Chinese economy. In retrospect, it appears that for some Chinese leaders this was interpreted as the successful completion of the socialist revolution. Further change toward a Communist society would be contingent on further transformation of the productive forces during an indefinite period of the future. Accordingly, the socialist state should take the lead in mobilizing the necessary expertise and resources to change the productive forces. To oversimplify somewhat, according to this view, technological advance would be the agent for further social changes.

For Mao and his followers, however, it seems that the socialist transformation of the economy only indicated the resolution of *antagonistic* contradictions within Chinese society. It is highly significant that at the time when socialist transformation was being celebrated, Mao prepared his important "On the Correct Handling of Contradictions Among the People." This work pointed to the persistence of non-antagonistic contradictions that had to be resolved lest they become transformed into antagonistic contradictions, a transformation seen as opening the possibility for a restoration of capitalism. Thus for Mao, the revolution in the relations among men had to continue. Further development of productive forces without further revolution would exacerbate the contradictions in Chinese society. Ultimately, without further change in the relations of production and the superstructure, the development of the productive forces would reach an upper limit. From the Maoist point of view, that upper limit would be reached quickly unless the creativity of the masses was unleashed. The full potential of the masses could only be realized by resolving contradictions and maintaining the primacy of politics over a more strictly economic approach to developing productive forces.[18] Thus, changes in the productive forces (technological advances) were dependent on further changes in the relations of production and the superstructure. This implied a strategy for technological mobilization characterized by mass activism aimed both at change in the relations of production and at technical innovation. Whereas the non-Maoist approach came to be criticized as "mechanization without revolutionization," the Maoist position can be thought of as "revolutionization for mechanization." The technological revolution, which accompanied the Great Leap Forward, marked the first serious attempt at implementing the Maoist socio-political approach to technological mobilization.

Technological Mobilization During the Great Leap Forward

The Great Leap Forward Years, 1958-1961, contained three significant developments that bore directly on efforts at technological mobilization. First, ideological considerations had a much greater role in national policy for scientific and technological development than previously, and in attempting to implement the

Maoist approach noted above, technological mobilization took the form of a mass movement. Second, technological mobilization was given a more integrated organizational backdrop with the centralizing of science policy into the hands of the State Scientific and Technological Commission (SSTC) and with the merger of the Association for the Dissemination of Scientific and Technical Knowledge with the All-China Federation of Scientific Societies to form the Scientific and Technical Association (STA). Finally, the economy was decentralized, and the management of production enterprises was changed.

Ideology

Chinese ideology contained a number of prescriptions for the development of science and technology and for the role of science in a socialist society. During the Great Leap Forward these were made explicit and were literally interpreted to form a set of prescriptions from which policy could be inferred.

As we have seen, Chinese ideology contains an implicit historiography of science that we have called "externalist." According to this view, the development of modern science is explained by economic and social factors outside science itself, rather than by the "internal" evolution of scientific ideas. This view emphasizes the historical contributions of nameless tradesmen and innovative workingmen rather than famous scientists or natural philosophers. In capitalist societies, these contributions are seen as having gone unrecognized and were either suppressed or stolen from the working class for use in advancing the economic and political interests of the exploiting class.

Ideology also contained epistemological principles that related to science. Foremost among these was the doctrine of the unity of theory and practice. Knowledge originated in the practice of production. Production experience is generalized into theory, which is then subject to verification in practice.

Finally, there are elements of what we have called "cultural nationalism" in Chinese ideology.[19] The Chinese people are seen as a great people with a great tradition and as a people who are most decidedly not inferior in science. China's state of relative underdevelopment in science is attributed to an entrenched system of feudalism that frustrated her scientific genius.

The implications drawn from ideology had direct relevance to policy. Historiography directed attention to the environment for science in China. With the proper material support and organization, and under Party leadership, science was expected to flourish. The notion of organization was important since scientific advance was viewed not as the product of isolated genius but of collective activity. It was therefore important to structure creative interaction to take full advantage of collective wisdom.

The idea of the unity of theory and practice meant that divisions between theorists and practitioners had to be overcome. Theorists—professional scientists and technologists—had to be brought in contact with the masses, and learn from and cooperate with them. Science had to serve production not only to increase production, but also to test the validity of science.

Since a socialist society was in the process of being realized in China, ideology led to expectations of a flowering of mass creativity. It would be important, therefore, to remove any residual obstacles to mass science. Workers and peasants should be encouraged to innovate and experiment, they should be provided with necessary scientific knowledge and resources, and they should be rewarded with acclaim for their contributions.

Finally, cultural nationalism meant that efforts should be made to record, digest and use the scientific and technical accomplishments of Chinese tradition and to emphasize indigenous, as opposed to foreign, capabilities. The best known of these efforts has been in medicine, but during the "Great Leap" attention was also given to recorded and accumulated information in geology, meteorology, botany, and agriculture.[20]

To implement these ideological themes, the Chinese reverted on a mass scale to the political mobilization techniques used during the pre-1949 revolutionary period and in subsequent political campaigns. These techniques involved a vibrant Party organization capable of leading groups in meetings for criticism and self-criticism and of arousing mass consciousness and emotions by identifying an object (or enemy) to be criticized and struggled against.[21] In the technological mobilization efforts of the "Great Leap," the objects included nature itself, but more importantly, they included what the Party thought China's "bourgeois experts"—scientists, engineers, and managers—represented. These bourgeois experts were portrayed as being divorced from production and as being excessively preoccupied with book knowledge drawn from foreign experience. They were therefore at odds with two tenets of ideology. Using the bourgeois experts and what they represented as targets to be struggled against, the masses could be aroused to act and take initiative in the interests of both improving indigenous Chinese technology and adapting imported technology to China's needs and conditions.[22] At the same time, the bourgeois experts were subject to ideological reform sessions directed at making their professional ethos, hence their work, more relevant to China's needs.

The significance of using political mobilization techniques for technical mobilization is difficult to assess. It would seem that such techniques are intended to disrupt the operating procedures and structure of authority, particularly in technical matters, in a given work environment. This disruption would cause organizational indeterminacy and personal uncertainty among all the participants—workers as well as managers and engineers. One possible effect of this disruption would have been that if, on certain given projects, worker rebellion against the technical staff on an issue of innovation did result in an objective advance in technology, worker self-confidence would have been greatly enhanced. Ideology, with its prediction of worker and peasant wisdom, would have received a measure of confirmation and a significant social-psychological change will have occurred.

On the other hand, criticism of technical experts and disregard for their advice could have and probably did lead to serious technical mistakes. These in turn would have led either to a reinforcement of feelings of inferiority in the face of technical authority or to a realization by the workers that they needed further technical training. The role of the cadre would then become crucial in channeling disappointment with mistakes into the latter, more productive course.

To summarize then, the use of political mobilization techniques for technological mobilization has real potential for achieving social change conducive to the creation of an innovative society. This is true both for changing the formal organization and procedures of a work environment and for effecting the kind of social-psychological change that would make the formal organizational change persist. The use of such techniques, however, also holds the potential for regressive, anti-developmental change in the face of failure. To further explicate the use and effects of these techniques, it will be useful to examine the broader context of technological mobilization during the "Great Leap."

Organizational Adjustment for Technological Mobilization

Complementing the themes expressed in ideology were a variety of organizational changes during the "Great Leap" period designed to take Chinese society on a new path to development. First, the Chinese economy was to be decentralized, and as we have seen, the science system was adjusted accordingly. In a Chinese- or Soviet-style economy, decentralization requires workable coordinating mechanisms in the subnational units that have responsibility for administration.[23] In China (and the Soviet Union) the Communist Party has performed this function. In China, decentralization has usually meant the strengthening of the power and influence of the Party in the provinces at the expense of the government apparatus, particularly the central government ministries.[24] Decentralization during the "Great Leap" was also accompanied by a change in management practices in enterprises. The one-man accountability system of the earlier period was replaced by management committees, which included worker participation and a leading role for the local Party representatives.[25]

The net effect of these changes was that power shifted away from technically oriented managers, with their strictly economic point of view and concern for fulfilling the quotas of the plan with quality products, to Party cadres who were more sensitive to the implications of ideology with its broader socio-political concerns. Where the reward structure of the manager was geared to the fulfillment of production quotas, that of the cadre during the Great Leap Forward was geared both to quotas and the implementation of ideological injunctions. The result was that the organizational environment loosened up and authority was in the hands of those who would encourage the masses to participate in research, bridge the gap between professional science and production, and implement with vigor the movement for worker and peasant innovations.[26]

In the science system itself, there was a simultaneous centralization and decentralization. Since 1957, there had been a steady drift of authority over all scientific matters to the centralized Science Planning Committee. This trend culminated in the merger in 1958 of the SPC with the State Technological Commission to form the State Scientific and Technological Commission, the supreme body for science and technology affairs until the Cultural Revolution. In compliance with the national decentralization policy, however, the SPC ordered the establishment of a regionalized research system in 1958. This order entailed the establishment of provincial branches of the Academy and of the Academy of Agricultural Sciences, which was established in 1957 as well, and the establishment of science and technology committees at provincial and subprovincial levels. Both the branch academies and the committees were to be under provincial Party control, although they were ultimately responsible to the center. The committees were to draft provincial science plans and have overall responsibility for the scientific and technological activities below them, much as the SSTC had responsibility for national activities.

A second organizational innovation of 1958 was the merger of the Association for the Dissemination of Scientific and Technical Knowledge with the All-China Federation of Scientific Societies to form the Scientific and Technical Association. The STA's broad mandate was to "develop the characteristics of combining scientific technology and mass organization to promote a technical revolution."[27] Although the control of the Association rested with the national organization—a National Committee, Presidium, and Secretariat—by its very nature, it was to have extensive field operations. Thus, local associations were to be organized in every province, city, special autonomous area, and county. The local branches were to be directed by the local Party organization and to receive technical supervision from the next higher level of the Association.[28] In addition, the STA was to set up branches in mines, industries, people's communes, and schools—again with local direction by the Party organization. Local branches of the scientific societies were to be established under the direction of the STA and the local political authorities, with technical advice and supervision coming from the national organizations.[29]

With the emergence of branch academies and local-level science and technology committees, the establishment of local STAs was very much a part of the decentralizing of a science network to bring research closer to production. At the lowest level of the network was the research group that existed at a low level of production, such as a factory, production brigade, or even a production team. The network also included STA-supported spare-time colleges for workers and peasants. A 1960 account reported that eight such colleges were opened by scientific societies under the STA in Shanghai, seven were opened in Harbin, and societies in Wuhan were also planning this kind of work. In each case, it was the scientific societies working under the direction of the STAs that provided the specialized competence.[30] In addition the societies continued to provide other kinds of technical support to the STAs through lectures, films, broadcasts and the activities centered around the STA-sponsored "halls of science" in the major cities.

The combination of the ideological atmosphere and the organizational changes that took place during the "Great Leap" resulted in an approach to technological mobilization that incorporated some of the features of the earlier approaches but with the intensity of a mass movement. Worker and peasant innovation campaigns were widespread and numerous techniques were employed to bring trained scientists and engineers into contact with the masses, and with problems of production.

An account of the evolution of technological mobilization in one rural area may be taken as an illustration of how the system was supposed to work. Many of the elements of technological mobilization discussed above are evident in this report. The account centers around Wang Pao-ching, a technical activist in Lichuan county, Shensi province. As early as 1952, Wang proposed that greatly increased maize yields could be realized with the introduction of scientific farming. His neighbors, who did not share his appreciation of a scientific approach, ridiculed Wang's statements. Wang's enthusiasm was spotted by the local Party representatives, and when Wang approached the Party for assistance, the Party arranged to send a team of agro-technicians to the village to introduce new methods. These methods were tried experimentally and were successful. The trials attracted much attention, and their success had a powerful demonstration effect. Appreciation of the scientific approach and Wang's "can do" spirit spread, and allegedly created a demand for technical information.

By 1955, Wang's cooperative had established a relationship of technical cooperation with the Northwest Agricultural College in the city of Sian. Professors from the College frequently came to the cooperative to sum up the experience of the farmers and help direct experiments. Wang's impact was noted and supported by the Shensi provincial Party committee, which arranged further support for the experimental work of the cooperative.

When Wang's cooperative, along with 35 others, was consolidated into a commune, the experimental work was enlarged. The commune set up an agricultural technical school and research committee. Each of 17 production brigades had its own primitive laboratory, below which were 121 more specialized teams that experimented with new techniques on specially designated plots. In addition, the commune set aside a piece of land as a "miracle experimental plot" where anyone could try out a novel idea. Wang's own lot in life improved as well. He became the head of the commune, was sent to study at the Agricultural College in Sian, and was made a special research fellow at the Shensi Agricultural Research Institute.[31]

Post-"Great Leap" Adjustments

The period of retrenchment that began in the early 1960s as a consequence of the Great Leap Forward actually began to check the mass movement approach to technological mobilization a bit earlier. At a STC-convened meeting of local science and technology committee representatives in May 1959, there was a call for consolidation, more rational planning, and better utilization of technical and material resources.[32] Some of the problems with technological mobilization during the "Great Leap" may be inferred from a worker's discussion of the technological innovation efforts in his factory.

According to this account, the collective, or team management practices that began during the Great Leap Forward stimulated worker innovations. The new managerial arrangements made it possible to circumvent the earlier regulations governing worker innovations and rationalization suggestions that called for a careful review of all innovations by engineering departments before adoption. This review process was seen as a serious bottleneck since there were not enough engineers to handle the load of proposals. The new managerial procedures eased the situation, but once widespread worker participation in technical matters began, there developed a realization that a better grounding in science and technology would be necessary. Thus, the initial "Great Leap" technological mobilization emphasis shifted from mass innovations to the formation of experimental groups and spare-time education.[33]

In a 1960 report on the work of STA, the stress was more on the summing up of mass innovations and training of workers and peasants than on their creativity. The Shanghai branch of the Association alone had grown to include 42 academic societies and over 2,500 basic-level organizations with a membership of 100,000. The member societies took an active role in evaluating the innovations from the masses and would popularize the more promising ones at society-sponsored exhibitions.[34]

It is difficult to generalize for all of China about trends in technological mobilization. The basic-level experimental groups that were supposedly widespread during the "Great Leap" years are discussed in some accounts from 1964 and 1965 as essentially new phenomena. While the Shanghai STA could report extensive development by 1960, the provincial STA in Kwangtung was only being established in late 1961.[35] In one report of a conversation with a commune leader in North China, it was said that not much attention was paid to scientific experiments at the production brigade level until after the Cultural Revolution.[36] While this report may be true for this particular commune, and perhaps for others, it seems likely that for many others, brigade-level scientific work was taken seriously before the Cultural Revolution.

It is possible, however, to generalize about a shift in strategy that occurred in the aftermath of the Great Leap Forward. In industry, authority was again shifted to professional managers. Party participation in the running of enterprises was far greater than during the first five-year plan, but unlike the "Great Leap" years, the Party was far more interested in economic construction than in ideological themes. Worker innovations no longer took on the qualities of a mass movement, as the role of the engineering staff in the technical review of worker innovations was again emphasized.[37]

In agriculture, a series of bad harvests created a critical situation. The organizational excesses of the "Great Leap" were done away with and a policy that permitted the cultivation of private plots was instituted. The severity of the agricultural crisis probably resulted in the curtailment of the kind of low-level experimentation that had occurred earlier, although by 1963 and 1964, this was resumed. The high priority accorded agriculture in the 1960s resulted in a renewed drive for the technical transformation in that sector, but the approach was to rely on trained agro-scientists for the technical inputs.

In scientific research, as we have seen in the previous chapter, there was a new respect for the value of expertise and conventional research. Scientists' time was rationalized and they were released from an obligation to indiscriminately participate in production activities.

Technological mobilization continued, but the stress was on dissemination, popularization, and spare-time education. Exhibitions, lectures by representatives of scientific societies, popular science readers and children's books (one series entitled *One Hundred Thousand Whys*, which contained contributions from China's more eminent scientists, was particularly popular), and films were numerous and widespread. The role of STA was prominent in these activities.

The policy shift away from the mass movement approach of the "Great Leap" period did not mean that useful organizational innovations from the "Great Leap" were abandoned. A good example of the persistence of a "Great Leap" development is the network of STA-related organizations.

An insight into the way local STAs and their member societies had by the early 1960s penetrated into the production setting and linked up with basic level research and innovation groups is provided in a report by a Japanese traveler. The case involves the Shanghai branch of the Electronics Society, whose membership, as a result of "Great Leap" reforms, included workers as well as scientists and engineers. The Society provided a forum for worker members to discuss technical problems at manufacturing sites. Those problems that could not be solved by workers at the plant would be brought up to the next level of the Society. There the workers had opportunities to discuss the problems with "technicians, researchers, and scholars in order to obtain solutions to the problems."[38]

In 1962, the Party Central Committee called for stepped up agricultural research as well as scientific and technical assistance to agriculture. The implementation of this directive, particularly the question of assistance, required the search for a formula or formulae for implementation. One approach again was to utilize the resources of the STA. The Shanghai branch for instance, initiated a series of forums on agriculture in 1962. These grew out of investigations, conducted by the Association of the state of agriculture in the Shanghai area, which discovered a need for new knowledge in agriculture. The forum participants included commune Party secretaries and commune directors, some production brigade leaders and experienced peasants, cadres and technicians from county-level organizations and representatives of science and technology units from the city. The objective of the forums was to educate the rural participants in the scientific basis of agricultural practice. The main forum was to give rise to county-level STA forums to be conducted in cooperation with the technical organization at that level. One of the outcomes of these discussions was a recommendation to concentrate on utilizing young peasants, and primary and middle school graduates as agents for the diffusion of agro-science and technology. These would work through commune-run agricultural spare-time schools.[39]

The Academy of Agricultural Sciences, which was founded in 1957, had by 1964 extended its operations throughout the country. Tai Sung-en, the Deputy Secretary-General of the Academy reported that each of the 28 provinces and autonomous regions had its own research institute. The scientists associated with these institutes were working with state farms and communes on the outskirts of large cities and in high yield areas, which had been chosen as pilot units.[40]

The most ambitious program to implement the 1962 directive to improve agriculture was the establishment of demonstration fields. The demonstration procedure had been successfully used for some time but did not constitute a national program. The idea was to go beyond small experimental plots, where conditions could be carefully controlled, to much larger plots ranging up to 10,000 *mou* (1 *mou* = 1/6 acre), which reflected actual production conditions. The demonstration fields were to serve as links between experimentation and extension and popularization.[41]

By 1965, it was reported that one-fourth of all agricultural researchers were spending the major portion of their time on demonstration fields. The fields were run by the state, by counties, and by experimental groups. In Kiangsu province, for instance, a demonstration and experiment station was set up in one commune. Other communes, brigades, and teams then ran their own stations by emulating the successful results of the main field. Representative demonstration fields were to be selected for further development. These would then receive funds for procurement of modern equipment and facilities, and in this way the geographical distribution of agricultural research was to be broadened.[42]

A tension permeated technological mobilization efforts in the early 1960s because of the split in the Chinese leadership on the correct path to technological development. With the legitimization of academic research in the immediate post-"Great Leap" years, the deployment of technical manpower was constrained. Technical personnel could legitimately claim that their time would be best spent in laboratories or in agro-science on experimental farms attached to research institutes. The elevation of expertise meant that the research results and new techniques developed in the research environment would have to be transferred via demonstration fields and extension mechanisms. Thus, an element of differentiation, reminiscent of the pre-"Great Leap" period, was introduced which had a compartmentalizing effect on technological mobilization.[43] Throughout the early 1960s however, there was also an undercurrent of opinion that favored the more direct involvement of scientists in production and opposed the organizational differentiation that allegedly created artificial boundaries between research and production.

A *People's Daily* account of July 1963 presented one side. While it praised the cooperation that had developed between a research institute and the county and subcounty science and technology organizations, it basically upheld the principle of differentiation noted above. The importance of well-equipped laboratories and laboratory research was asserted, and it added that

> . . . scientific workers can only participate in the popularization work which covers the subject of their research work. . . . They participate in the popularization work in order to carry out their scientific experiments but not to perform the routine work of an agro-technique propaganda station.[44]

In contrast to this position is a carefully articulated argument for agricultural researchers to be more involved with production. Yü Ch'i-pao, Director of the Northwest Research Institute of Agricultural Sciences, Ministry of Agriculture, writing in the November 1963 issue of *Red Flag*, set down five stages for cooperation between agricultural researchers and the masses. First, research topics should be aimed at rapid agro-technical reform, sustained increases in yield, and the raising of labor productivity. Second, Yü addressed himself to the question of the location of research work. He noted that this depends upon conditions but that the principle of the closer to production the better should be employed.

The divergence with the previous view becomes clearer in Yü's discussion of the third stage, demonstration. He argued that it is not sufficient for agronomists to produce research results. "The masses can hardly appreciate a new technology without personally testing it from beginning to end." Therefore, those most familiar with the research should undertake propaganda work. Yü noted that new innovations are only successful after a complete work cycle had been completed with agronomists and the masses working closely together. According to Yü, ". . . it is absolutely necessary for the creator of new research results to take part in their demonstration." This work was to be accompanied by scientific and technical education to raise the technical level of the masses.

Popularization is the fourth stage and involves spreading successfully adopted new techniques to a wider area, involving more people. Again, the participation of the expert is regarded as essential, since conditions change from area to area. The expert is to be available and participate in the adaptations to local conditions that would be necessary. Yü argued that without this involvement, popularization cannot be fully successful and the initial group who adopted the technique would become an "isolated army."

Finally, the spread of an innovation must be monitored and evaluated to discover secondary and tertiary effects. Yü wrote:

Prior to the introduction of the results of scientific research, local production, people's living conditions and biological colonies (*sheng-wu ch'ün lo*) are in a definite state of balance and good order. The introduction of an important result of scientific research will upset the balance, and a new balance must be established within a definite period of time. . . . Only in this way can the new results of scientific research be consolidated.[45]

By 1964 there were signs that technological mobilization was again assuming the form of a mass movement. Peasant research groups were again emphasized; they were to be governed by the mass line ("from the masses, to the masses") and reflect the ideal triple combination of basic-level cadres, veteran peasants, and educated youth. Their objective was to achieve a "close combination of modern science and technology with the Chinese peasants' traditional experience." The value of peasant wisdom was again being praised and rewarded. As one example, peasant wheat expert Liu Ying-hsiang was provided a laboratory by the Honan Province science and technology committee and was appointed a corresponding research fellow of the Honan Provincial Agricultural Research Institute.[46] Thus, a subtle shift was occurring in the relationship between higher technical competence and local producers–from one of direction from above to one of support from above.

The basic level research groups of cadres, veteran peasants and educated youth were celebrated as "a good form of organization for developing more scientific experimental activities in rural areas."[47] The importance of Party leadership through local CCP committees was suggested in that the development of these groups "would necessarily involve the ideological struggle between the new and the old, the progressive and the conservative, personal interests and collective interests, and the struggle between two roads." However, it was also recognized that the level of scientific and technical knowledge available in the groups would be limited. The STAs, particularly at the county level, were to provide the kind of technical support needed to enable the basic level groups to operate.[48]

At the same time the political as well as the practical significance of technological mobilization was noted, and the mass nature of STA as well as its technical nature was emphasized. As the *People's Daily* editorialized:

Being expert and mass organizations, these science and technology associations are well-qualified to implement the mass line in science and technology activities and to develop the role as a connection between units at different levels and between parallel units. Through special academic societies and groups, the science and technology associations may in accordance with different conditions organize the scientists and technicians appropriately according to their specialties and needs to carry out such activities as technical inspection, summing up of experiences, discussion of problems, and teaching of techniques.

On the one hand this will disseminate modern scientific and technical knowledge and give technical guidance over mass scientific experiments held at the basic level; on the other hand, it may familiarize the members of science and technology organizations better with the experiences acquired by the masses and the problems met in production, and promote the improvement of scientific work.[49]

Technological mobilization in industry went through a process similar to agriculture: from retrenchment in the immediate post-"Great Leap" period, to gradual re-emphasis on mass participation. Three aspects of this evolution are worth noting here.

First, after the initial shift of authority back to the professional managers, there developed concern for both the political and practical implications of this structure of authority. The concern for standardization and product quality, efficiency, and imported technology tended to enhance the position of the professionals, and certainly inhibited worker innovations. At the same time, the same complaints about compartmentalized expertise being somewhat irrelevant to actual production that had surfaced in agriculture were heard in industry. These developments were addressed by Mao in November 1964 in his call for a revolution in designing, which was intended to overcome the political and practical difficulties discussed above.[50]

A second development during this period, related to the authority question in industry, was the emergence of 3-in-1 groups in the industrial sector. These consisted of workers, cadres, and technical personnel, and although they first appeared during the Great Leap Forward, the shift in management in the immediate post-"Leap" period may have resulted in the eclipse of these 3-in-1 groups. However, by 1964-1965, it was reported that the 3-in-1 formula had been adopted by industrial enterprises throughout China.[51] In the pre-Cultural Revolution period, the 3-in-1 groups appear to be far more involved in communication and education than in management and technical decision making. This would change as a result of the Cultural Revolution.

One development that was new to the 1963-1965 period was the mass technical cooperation movement. The movement began in Shenyang when a worker came up with a labor-saving innovation. Although government and plant-run mechanisms for popularization of this kind of innovation existed, they were often slow and did not deal with day-to-day problems. Therefore, these workers took the initiative and reportedly traveled around from plant to plant in the Shenyang area on their bicycles proposing the establishment of an informal interplant technical consultation service. After meeting resistance from some fellow workers, who were either too busy or felt the activists were showing off, they finally succeeded in organizing a small group. At this point, their efforts were recognized by the Shenyang Party committee and the Liaoning Trade Union Council. The efforts were formalized, and rules of procedure for utilizing the consultation service were established.[52]

A final point about technological mobilization in the period between the Great Leap Forward and the Cultural Revolution concerns the extent to which China relied upon the importation of foreign technology. The "Great Leap" period was characterized by a great stress on native methods and indigenous skills. This was during a period when foreign technology was still readily available through cooperation with the Soviet Union. After the "Great Leap" and the withdrawal of Soviet assistance, there appears to have been a renewed emphasis on foreign technology from whatever sources were available. This is explained by the Chinese and by Western observers as a result of Liu Shao-ch'i's growth in influence following the Great Leap Forward and Mao's relative decline in influence. During the Cultural Revolution, the Liuists were accused of being excessively dependent on foreign technology and of failing to realize and encourage the indigenous capabilities of the Chinese people.[53]

This allegedly Liuist "enslavement" spread throughout the nation's research and designing activities and was viewed not only as ideologically inappropriate but technically irrational since it did not permit the modification and adaptation of foreign technology to Chinese conditions. Thus, in one case offered as an example of the pre-Cultural Revolution conservatism of bourgeois experts, workers who proposed improvements on a piece of imported machinery were informed that "This is a foreign machine. Every single part of it is based upon a scientific theory."[54]

The significance of this orientation for technological mobilization is that, had it been allowed to continue, it may have placed China on a course, quite comparable to other low-income countries, of being excessively dependent—materially and psychologically—on foreign sources of technology, with all the adverse consequences that reliance entails for the development of indigenous research and development.

Developments Since the Beginning of the Cultural Revolution

During the Cultural Revolution, this dependent orientation was soundly criticized, and the contributions of China's masses were again praised. The foreign versus indigenous controversy was not a simply defined Maoist-Liuist policy difference on technology. It also involved foreign policy, fiscal and monetary policy, and questions of social policy that have implications for the course of the Chinese revolution. The remarkable interest in advanced technology from abroad shown by the Maoists in recent years, after the Liuists were eliminated, demonstrates that the issue is not clear cut. Rather the differences seem to have involved nuances and interpretation, and questions of selectivity and discrimination. Defense-related technology for instance was an area in which the Maoists felt no compromise could be made on the question of self-sufficiency. The Liuists on the other hand, apparently were willing to compromise on this issue, particularly during the early years of the build-up of American forces in Vietnam. Another area where interpretation was important was in dealing with China's more experienced scientists and engineers. These people generally were trained in foreign countries and the relatively free exercise of their expertise, advocated by the Liuists, entailed that essentially foreign solutions would be offered to China's technical problems. While in many instances "foreign solutions" may have been universal solutions and indeed the best solutions, this approach was not entirely consistent with the Maoist vision of the growth of an indigenous scientific tradition in which modern "foreign" science would be integrated with traditional Chinese wisdom and the contributions of the masses. "Make things foreign serve things Chinese" has emerged as the correct Maoist line.

Thus, the course of the Cultural Revolution involved removing obstacles to the institutionalization of the Maoist approach to technological mobilization, with its overriding objective of insuring that "the working class would become the true masters of science and technology."

The changes in approaches to technological mobilization that occurred during and since the Cultural Revolution have been designed to insure that the Maoist approach would predominate. One 1968 report summed up the situation nicely. The Liuists were alleged to have stressed that the principal contradictions in Chinese society were between the need for machines and a shortage in productive capacity. This led to programs of "mechanization without revolutionization." The Liuists allegedly did not trust the working class, and therefore ignored the essential question of whether, in industrial and business management, power was in the hands of the working class. Instead, under the guise of "scientific management," power was shifted to the hands of "bourgeois experts."[55]

The organizational changes that occurred were not in the direction of founding new organizations and programs for technological mobilization but rather seem directed to rectifying the operation of existing organizations or returning to the organizational innovations of the Great Leap Forward. More than one report states that worker research groups that grew up in factories during the "Great Leap" were either shut down or frustrated on orders from the Liuists.[56]

The 3-in-1 formula was maintained and elevated to greater significance. No longer was it merely a communication mechanism, with management playing the leading role. Now the voice of the worker was to carry equal weight; the initiative of the worker would predominate while management and technical experts would support the worker efforts. It should be noted parenthetically that the Maoists have always been aware of the low technical level of the masses and for the need for technical inputs from those with more advanced training. The question has been, under what terms is this to be done. The Cultural Revolution reforms were to insure that the terms were those acceptable to the workers and consistent with their production needs.

The 3-in-1 formula that is now widely used for technological mobilization in both industry and agriculture appears to be an empirical expression of the conditions for technological mobilization set forth analytically at the outset. The technical personnel serve as channels for the kind of information necessary for the process. The managerial and Party cadres have access to the necessary resources and the worker and peasant masses are both the object of the mobilization, and the condition for its success. The masses must be energized to appreciate and desire technological innovation. A kind of technological demand must be created. The Chinese have felt that this is best accomplished when the masses can identify with innovations and can develop confidence in the innovations they themselves participate in. The problem of leadership has been how to best demonstrate that it is in the interests of the masses to respond positively to new technologies and scientific knowledge. The Cultural Revolution reforms were designed to give workers and peasants a greater voice in expressing those interests.

Chinese experience has shown that the methods for technological mobilization have alternated over a period of twenty years between what might be called the economic approach versus the socio-political approach. The important question has been what kinds of priorities has the central Party leadership set for the society and what kinds of approaches are used to win compliance from the citizenry. During the First Five-Year Plan and the 1961-1965 periods the economic approach predominated. Because the Party center was committed to modernization that would be measured by short-term economic performance, and Party cadres would be evaluated by the economic performance of the production units they were assigned to, they naturally allied themselves with technical experts—managers, engineers and designers, and scientists—who they thought could help them deliver. The stress would be on technological mobilization from the top down. Resources would be carefully managed and where a question arose about investing in production inputs with immediate returns or in worker or peasant research and innovation, the former would be chosen.

The socio-political approach was predominant during the Great Leap Forward and since the Cultural Revolution. It is against neither economic development nor advanced technology, but it offers an alternative approach to economic development by integrating advanced technology with present Chinese conditions while at the same time urging the development of indigenous intermediate technology. Ideology, which serves as a blueprint for integrating economic, social and political development, receives highest priority at the Party center, and cadre performance is judged by his achievement of ideological not merely economic goals. Since ideology stresses the importance of the technical transformation of the masses and the contributions of the masses to the technical transformation of the society, the choice between short-term economic gains and the achievement of socio-political goals will be decided in favor of the latter. Technological mobilization is from the bottom up.

The two approaches have their unique costs as well as their unique benefits. The economic approach has obvious socio-political costs. As organizations are differentiated by specialization and technical expertise is given a relatively free hand, communication becomes a problem, undesirable social stratification develops, the prospects of technological alienation are enhanced, and a vibrant Party organization becomes ossified as it forfeits its leadership prerogatives to the experts. As we have seen, the purely economic approach may have practical costs as well, since it does not fully achieve the diffusion and adoption of new technology on which it is premised, precisely because that diffusion and adoption are not simply economic problems.

The costs of the socio-political approach can be measured in terms of wasted or insufficiently used human and material resources. Trained technical personnel who must spend too much of their time in activities that others who lack their special qualities could do just as well is wasteful of a scarce resource. Similarly, the diversion of resources for worker and peasant research and innovation means that resources will not be available for investment in projects where desirable results are more predictable. The kind of loose, unregularized social milieu that results from emphasis on the socio-political approach may stimulate innovation, but it may hurt production since there is a greater likelihood that product quality will decline, and since the informational and procedural requirements of a planned economy are sacrificed. On the other hand, the regularized and consolidated milieu resulting from the application of the economic approach is probably conducive to, and indeed necessary in, a planned economy but may seriously hamper innovation.

Chinese experience must be seen as a result of the tensions that exist between these two approaches. Excesses inherent in one approach cause a reaction that leads to the employment of the other approach. When excesses develop there, the pendulum swings back. In keeping with the pendulum swing, the period since the end of 1971 has seen a slow movement back to the economic approach, in response to the excesses of the Cultural Revolution and immediate post-Cultural Revolution periods. Events since the beginning of 1974 suggest that the swing towards the economic approach may have been arrested and that movement towards the socio-political approach may be beginning again. However, in these shifts in emphasis, there does appear to be a learning process going on. Successful modes of technological innovation persist. Workers and peasants who have been partially mobilized during periods of socio-political emphasis are better prepared for the economic construction tasks which are emphasized by the economic approach. Similarly, the value of expertise, demonstrated during the periods when the economic approach is taken, may lead to less inefficiency and waste during periods of socio-political emphasis.

How long these alternations will continue in China is impossible to predict. It can be said, however, that they have been a unique, and somewhat attractive method of creating a national tradition of interest in science and an appreciation of the value of science and technical innovation for production.

Notes

1. This concern was nicely expressed by one-time Vice-Premier Li Fu-chun.

> If scientific and technical levels, particularly of the working class, peasantry, and intellectuals, could not rise in step with this development (of economic production), contradictions will grow between the expanding productive forces and the ability of the men who control them.

"Li Fu-chun on Importance of Popularizing Science," *New China News Agency (NCNA)*, October 30, 1950, in U.S. Consulate General, Hong Kong, *Survey of China Mainland Press (SCMP)* 1403.

2. Karl W. Deutsch, "Social Mobilization and Political Development," *American Political Science Review*, LV (September 1961).

3. See Hahn-Been Lee, "Developmentalist Time, Developmental Entrepreneurs, and Leadership in Developing Countries," in Dwight Waldo (ed.), *Temporal Dimensions of Development Administration* (Durham: Duke University Press, 1970).

4. See Everett M. Rogers and F. Floyd Shoemaker, *Communication of Innovations: A Cross-Cultural Approach* (New York: The Free Press, 1971), Chapters 9 and 10.

5. See Joseph S. Berliner, "Bureaucratic Conservatism and Creativity in the Soviet Economy," in Fred W. Riggs (ed.), *Frontiers of Development Administration* (Durham: Duke University Press, 1970), pp. 586-592.

6. At present, there seem to be two approaches to understanding technical innovation campaigns in China. Renssalaer Lee has studied them from the point of view of a political scientist interested in the role of ideology, nationalism, and political leadership in technological development. See, his "Ideology and Technical Innovation in Chinese Industry, 1949-1971," *Asian Survey* XII 8 (August 1972), pp. 647-661.

Genevieve Dean, on the other hand, has viewed the campaigns as an alternative source (other than R&D) of technological innovation for Chinese industry. See her "Science, Technology and Development: China as a 'Case Study,' " *The China Quarterly* 51 (July-September 1972), pp. 520-534; and "A Note on the Sources of Technical Innovations in the People's Republic of China," *The Journal of Development Studies* IX 1 (October 1972), pp. 187-199.

7. Rensselaer Lee, op. cit.

8. Ibid., p. 650.

9. Cf. Ibid,, pp. 648-651, and Berliner, op. cit.

10. Just as the political activist, as a link between the Party and the masses, is a key element in political campaigns, so the technical activist is a key element in mass innovation campaigns. For both the political and the technical, the activist embodies important values which the regime hopes to propagate. In both cases these include a willingness to defy and struggle against some condition or authority ("bourgeois" and foreign science and technology and the capriciousness of nature in the case of the technical activist), initiative, perseverance, optimism, etc. In short, most innovation campaigns and the use of technical activists are of the same genre as the better-known political campaign style of leadership. For a discussion of political activism, see Richard Solomon, "On Activism and Activists," *The China Quarterly* 39 (July-September 1969), pp. 76-114.

11. It is important to distinguish between the historical meaning of "entrepreneur," which includes ownership of an enterprise as one of its defining characteristics, and the looser meaning used here. Usage here is intended to refer to the traits of the risk-taking innovator that can be found in individuals in capitalist and socialist systems who are not necessarily owners of enterprises. (*Cf.* Chapter 1.) These same traits have recently been associated with what one author has called a "product champion," an innovator in a firm who has the imagination, perseverance, and "ego-riskiness" to champion a new idea and see that it is implemented. See Alok K. Chakrabarti, *The Effects of Techno-economic and Organizational Factors on Adoption of NASA Innovations by Commercial Firms in the United States*, unpublished Ph.D. dissertation (Northwestern University, 1972), Chapter 8.

12. Many experienced workers in China have been identified as activists and have been promoted to "technician" and "engineer" positions. Indeed, this procedure has been an important means for meeting China's technical manpower needs. See Chu-yüan Cheng, *Scientific and Engineering Manpower in Communist China, 1949-1963* (Washington: National Science Foundation, 1965), pp. 61-62.

13. Kuei-sheng Chang. "Nuclei-Formation of Communist China's Iron and Steel Industry," *Annals of the Association of American Geographers* 60 (June 1970), p. 275ff.

14. *China Reconstructs* (November 1955), p. 14.

15. See Rogers and Shoemaker, op. cit., p. 14.

16. "Strengthen the Work of the Popularization of Scientific Techniques to Insure the Fulfillment of the First Five Year Plan," *Kuang-ming Jih-pao (KMJP)*, August 26, 1955, in *SCMP* 1129.

17. "First National Conference of Activists in the Dissemination of Scientific and Technical Knowledge Opens," *NCNA*, October 29, 1956, in *SCMP* 1412.

18. For a recent statement supporting this analysis, see Ch'uan Chung, "Criticize the Reactionary Essence of 'The Theory That Only Productive Forces Count' Advocated by Lin Piao," *Kuang-ming Jih-pao*, December 17, 1973, in *SCMP* 5535.

19. See footnote, Table 4-2.

20. For agriculture, see Leslie T.C. Kuo, *The Technical Transformation of Agriculture in Communist China* (New York: Praeger Publishers, 1972).

21. For a summary of these techniques, see Frederick Yu, "With Banners and Drums: The Mass Campaign in China's Drive for Development," *Current Scene* IV 9 (May 1, 1966).

22. For a discussion of attitudes toward foreign technology and the objectives of the innovation campaigns of the "Great Leap" period, see Rensselaer W. Lee III, "The Politics of Technology in Communist China," in Chalmers Johnson (ed.), *Ideology and Politics in Contemporary China* (Seattle: University of Washington Press, 1973), pp. 301-325.

23. For a good discussion of this question, see Jerry Hough, "The Prerequisites of Areal Deconcentration: The Soviet Experience," in James J. Heaphey, *Spatial Dimensions of Development Administration* (Durham, North Carolina: Duke University Press, 1971), pp. 132-175.

24. Franz Schurmann, *Ideology and Organization in Communist China* (Berkeley: University of California Press, 1970), Chapters II and III.

25. Ibid., Chapter IV. See also Stephen Andors, "Revolution and Modernization: Man and Machine in Industrializing Society, the Chinese Case," in Edward Friedman and Mark Selden, *America's Asia* (New York: Vintage Books, 1971), pp. 393-444.

26. For discussions of relevant aspects of this changed organizational environment, see Andors, op. cit., pp. 197-198; Audrey Donnithorne, *China's Economic System* (New York: Frederick A. Praeger, 1967), pp. 164-165; and Charles Hoffmann, *Work Incentive Practices and Policies in the People's Republic of China, 1953-1965* (Albany: State University of New York Press, 1967), pp. 95-102 and 115.

27. "Resolution on the Establishment of the Science and Technology Association," in U.S. Department of Commerce, *Joint Publication Research Service (JPRS)* 1502 DL.

28. Ibid.

29. Ibid. By 1959, it was reported that local STAs had been set up in 1,399 counties or municipalities with a combined membership of six million. *NCNA*, March 20, 1959.

30. See *NCNA*, March 14, 1960, in *SCMP* 2220; and *NCNA*, August 7, 1960, in *SCMP* 2316. Chu-yüan Cheng reports that by 1959, there were some 150 spare-time higher institutions in the country. By the end of 1960, 470,000 workers were reportedly enrolled in spare-time college, a figure which was approximately one-half the total number of full-time students enrolled in institutions of higher education at the same time. Cheng, op. cit., p. 58.

31. This account is taken from Chang Chi-tao, "How Wang Pao-ching Made His Farm a Laboratory," *China Reconstructs* (October 1960), pp. 32-34. Opportunities for upward mobility of technical activists, which Wang's case demonstrates, increased during the "Great Leap Forward" period. Chu-yüan Cheng

notes that approximately 10,000 workers rose to the rank of engineer during the years 1958-1960 and approximately 100,000 rose to the rank of technician. Cheng, op. cit., pp. 61-62.

32. *NCNA*, May 31, 1959, in *SCMP* 2037.

33. Fang Yu-ting, "We Are the Masters of Our Plant," *China Reconstructs* (March 1962), p. 8.

34. "Association of Science and Technology of Shanghai Sums Up and Popularizes Achievements in Technical Innovation," *Jen-min Jih-pao (JMJP)*, September 8, 1960, in *SCMP* 2340.

35. *NCNA*, November 18, 1961, in *SCMP* 2627.

36. A.Z.M. Obaidullah Khan, "Class Struggles in Yellow Sandhill Commune," *The China Quarterly* 51 (July-September 1972), p. 538.

37. See Fang Fu-lin, "What I've Learned in Eleven Years," *China Reconstructs* (June 1963), pp. 24-26. Fang notes that the right of decision in technical matters in factories lies with the chief engineer and chief designer, and the right of approval lies with the director, although many questions are solved through extensive discussions. *Cf.* Andors, op. cit., p. 416.

38. *Rajio Gizutsu (Radio Technology)*, March 1966, in *JPRS* 29, 280.

39. Chiang Chiu-chin, "Science and Technology Association of Shanghai Spreads Scientific and Technical Knowledge of Agriculture," *KMJP*, August 25, 1962, in *SCMP* 2823.

40. Tai Sung-en, "Directions of Agricultural Research," *China Reconstructs* (April 1964), pp. 14-17.

41. Kuo, op. cit., p. 23.

42. This discussion of demonstration fields is taken from Chin Shan-pao, "Demonstration Fields—Bases for Agricultural Research," *China Reconstructs* (June 1965), pp. 10-11.

43. For one discussion of the political significance of the demonstration fields, see Jan S. Prybyla, *The Political Economy of Communist China* (Scranton, Pa.: International Textbook Co., 1970), p. 441.

44. "Correct Way of Conducting Agricultural Scientific Experiments," *JMJP*, July 23, 1963, in *SCMP* 3037.

45. Yü Ch'i-pao, "Some Points Concerning Research to be Carried Out Jointly by Agronomists and the Masses," *Hung Ch'i* 22 (November 19, 1963), in *JPRS* 22, 660.

46. *NCNA*, September 7, 1964, in *SCMP* 3297.

47. "National Science and Technology Association Holds Meeting in Peking on Mass Experimental Activities," *NCNA*, May 20, 1964, in *SCMP* 3231.

48. Ibid.

49. "Steadily Develop Mass Scientific Experiment Activities in the Countryside," *JMJP* (editorial), May 21, 1964, in *SCMP* 3231.

50. See Rensselaer Lee, op. cit., pp. 657-658.

51. Ch'en Shih-ho, "Shanghai's Technical Revolution Scores New Victories," *China Reconstructs* (May 1965), p. 13.

52. Sun Ming-hui, "New Horizons for Innovators," *China Reconstructs* (May 1963), pp. 2-4. See also, *NCNA*, May 5, 1964, in *SCMP* 3215.

53. See Rensselaer Lee, op. cit., p. 657ff.

54. *NCNA*, November 20, 1968, in *SCMP* 4306. Cited in Rensselaer Lee, op. cit., p. 659.

55. "To Lead Mechanization with Revolutionization is the Broad, Level Road to the Development of the Machine Building Industry," *KMJP*, December 16, 1968, in *SCMP* 4332.

56. See, for instance, "There Are Good Prospects for Factories to Undertake Scientific Research," *KMJP*, February 28, 1971, in *SCMP* 4857; "Miners' Scientific Research Institute Achieves Important Successes," *NCNA*, April 6, 1971, in *SCMP* 4878; Cf. also Khan, op. cit.

6 Conclusion: On the Idea of a "Chinese Model"

. . . any intellectual or technical step has not only to be made once; it must become part of a tradition. Like a variant in organic evolution, it must not only occur–it must become established.

–S.E. Toulmin

Whether an ongoing, indigenous scientific tradition has been established in a nation cannot be easily ascertained over the short course of twenty years. It is, therefore, difficult to *evaluate* Chinese science policy and administration at this time. It may well be that future historians will regard the first twenty years of the People's Republic as a period of "essential preparation" for a flourishing of Chinese science and technology in the last quarter of the twentieth century that would thus catapult China into the position of being a world leader in science by the turn of the twenty-first century. Alternatively, apparent irrationality in Chinese policies and procedures may turn out indeed to be real irrationality and may have the effect of setting China back twenty to forty years and of making her dependent on foreign science and technology on a semi-permanent basis. In spite of the problems of evaluation, however, some tentative judgments can be made.

By a number of measures, Chinese performance has been impressive. Her successes with atomic weapons, missiles, and satellite development are significant. However, these are somewhat atypical, having been achieved under conditions of relative plenty and through organizational and administrative arrangements that were not subject to the twists and turns of ideology, political lines, and competing models. By more conventional and typical indicators, however, the progress is also laudable. The establishment of hundreds of research and development institutions across the country is quite impressive.[1] The commitment of an estimated 1.1 percent of GNP to research and development[2] and perhaps as much as 15 percent of government expenditures on science-related activities[3] places China with the world's big spenders on R&D and places her in a category of her own among the low-income countries. Similarly, the training of an estimated 1.3 million scientists and engineers in the 1949-1967 period is impressive, although that figure falls short of China's own goal of 2 million.[4]

Although these facts are important, they do not tell us a great deal about the more important, and more difficult to assess, efforts to create a society of producers who are positively oriented towards innovation and efforts to integrate R&D into that society. We have seen that one of China's goals was to use science to achieve technological innovation in production, and we have also noted the OECD finding that "technological innovation is more than R&D." A straightforward way to begin to evaluate China's progress in this regard, therefore, would be to measure research performance in terms of economic gains. Trying to measure research outputs in terms of production inputs is a formidable problem of conceptualization, however. The situation in China is even more difficult because of a lack of data for both research and production. If one had to guess, one suspects that China could have been marginally more successful in the short run in achieving this goal had she stuck to the bureaucratic-professional model that was emerging in 1957. However, it is important to stress the *marginal* and short-run nature of this improvement. It is possible that the bureaucratic-professional model would have evolved into a system that over the long run may have consumed many more resources than it would have contributed to producing. Such a system may have become highly routinized and bureaucratized, elitist, urban-based, and largely irrelevant to the technological needs of all but a few more sophisticated areas of the economy. This is what the Maoists have charged, and similar systems in countries like India have indeed demonstrated these characteristics.

On the other hand, the substitution of mobilization models was highly disruptive of professional and emerging administrative patterns. Such disruption undoubtedly had negative short-run economic consequences, and however much they were justified as increasing immediate relevance to production, it is likely that the disruptive costs outweighed the benefits.

The truly unique aspects of Chinese experience have been the attempts to integrate science and technology with social and cultural change. It is here that the mobilizational models are perhaps most relevant with their explicit attempts to break down barriers between intellectuals and the masses and between organizations in different sectors and to encourage and reward experimentation and innovation by workers and peasants. Any evaluation of these efforts to reverse the organizational differentiation between research and production and the growth of social stratification and social distance between educated scientists and engineers and the masses must begin with the question of evaluative criteria.

It is no easy task to set criteria for the establishment of a national tradition of science in the service of production. It is relatively easier to see how a tradition of pure science might be fostered. The emphasis for the latter would involve the selection and training of an intellectual elite, relatively high levels of funding for their research, and high degrees of autonomy in organizing their activities. Administration would be mainly concerned with insuring the proper supply of those material conditions required for the research. Evaluation could be based on publication output and citation analysis.

However, when one adds the requirement that research be relevant to national needs, the administrative and organizational problems grow ever more complex. Attention must now be focused not only on the research establishment itself, but also on the environment for research and the linkages between the research center and the environment. Here lie the differences in approach between the mobilizational and bureaucratic-professional models. The latter is based on the assumption that professional training qualifies certain members of society for special status and perquisites. The research results produced by this group are regarded as having a leading role in economic development as they somehow gradually filter down into production. Social transformation is regarded as a consequence of economic and technological developments, which force changes in social organizations. The bureaucratic-professional model assumes that the responsibilities of scientists and technologists are limited to teaching, research, development, and design work as well as some administration, and that research, development, and design activities are the special province of scientists and technologists. It presumes that a demand for new knowledge and technology exists in the society, and that responsibility for incorporating new technology in production belongs to others. As we have seen, however, it is precisely because there is no basis for this presumption in most low-income countries that extraordinary efforts must be made to create a technological demand and to link R&D with the economic activities of the society.

The mobilization models, on the other hand, see less justification for organizational differentiation and social stratification since innovation, and the application of science to production are regarded as functions of social organization. Social transformation is viewed as a prerequisite for economic and technological development. The objective, therefore, is to change social organization and the cultural level of the society so that the masses recognize, appreciate, produce, and demand scientific and technical knowledge and that research and innovation become a way of life for all. The mobilization models include as an objective the abolition of organizational barriers that create distinctions between research establishments and the environment by insuring an integration of production and research. In the absence of clear organizational procedures, leadership is less administrative than it is social-psychological, where ideology is used as an integrative system of symbols operating under conditions of institutional uncertainty.

There is no *a priori* basis for deciding which set of assumptions is most appropriate for the fostering of a scientific tradition in service to production. Reference to experience does not produce unambiguous evidence either. Those assumptions underlying the bureaucratic-professional model often seem more congenial to Western observers but may also be based on a biased view of Western scientific development, which ignores a developmental pattern of interrelated scientific, economic, and social factors. One can cite many cases where science led to technological development, which in turn led to social change, but as we have seen in Chapter 1, changes in social organization may also be prerequisites for technological change and scientific advance.

Here may lie the real significance of Chinese experience. The Chinese themselves have been unable to decide how science should be integrated with society and social change. As a result they have experimented with competing models. They have in the process both advanced indigenous research capabilities and also contributed to the creation of a society that appears to be innovation-oriented and prepared to utilize modern science and technology. Thus, the Chinese model for scientific development is, in fact, a series of competing models. One promotes research, education, specialized management, and innovation from the top down but tends towards social stratification and organizational rigidity; one promotes the integration of science with society in the interest of societal change but seriously neglects specialized training and specialized management and creates an atmosphere in which academic research as commonly understood is disrupted. The weaknesses and inconsistencies with Chinese goals inherent in one model produce pressures for change to an alternative model, which has its own strengths and weaknesses. The developments since the beginning of 1972, which point to changes away from mobilization model II back towards concern for structure, regularization, the special requirements for employing expertise, and the economic approach to technological mobilization, are entirely consistent with the perspective of competing models developed here as is the recent apparent swing back toward mobilization. Although it is still too early to characterize an emerging new model, it will probably reflect an accommodation between the mobilization II and the bureaucratic-professional models.

Implications of Chinese Scientific Development for the Administration of Change

Chinese organizational and administrative approaches to science programs epitomize more general administrative problems in China. A fundamental difficulty has been to reconcile leadership techniques that evolved over a long period of revolutionary struggle and governance of guerrilla bases with what some would call the organizational realities of modern states characterized by specialization and differentiation. Acquiescence to the latter entails the sacrifice of legitimizing values inherent in the former. Those values in turn achieved their salience in a situation of revolutionary struggle. A continuing issue in post-1949 China, of which the policy shifts in the science and technology area are but one manifestation, is whether the revolution continues or not. The position of the Maoists is that contradictions persist and the revolution is not complete, particularly the revolution of changing the social relations and culture of the Chinese people. Thus, those who would acquiesce to the "organizational realities of modern states" (the Liuists?) are in effect frustrating the revolution by making it possible for enemies of the revolution to reintroduce old patterns of authority and old patterns of thought.

In a very real sense, therefore, administrative development, which is tied to scientific and technological development, is also closely tied to the Maoist view of political development. Efficient administration, regularization of procedures, and rationalization of organizations without ideological growth and Chinese-style political participation is tantamount to premature institutionalization. Premature institutionalization in turn blocks the progress of revolutionary change and opens up the possibility of power being seized by "revisionists" and "capitalist roaders."

Foremost among the issues of political development are those concerned with attitudes toward authority and toward the resolution of conflict. Solomon has argued traditional Chinese socialization patterns produced two patterns of interpersonal behavior. In relations among peers, the suppression of conflict in the interest of harmony was considered a paramount objective. In superior-subordinate relations, fear of and deference to authority on the part of subordinates were the preponderant attitudes. These sentiments were used by traditional power holders to induce passivity in the population. Such passivity was consistent with the steady-state objectives of the traditional political-administrative order but would seem to be antithetical to China's twentieth century modernizing impulses. Modernization, generally, and the creation of a tradition of research and innovation, in particular, as we have seen, requires the opening up of channels of communication. However, fear of conflict in interpersonal relations had the effect of blocking horizontal communication among peers. Fear of authority had the effect of blocking vertical communications between superiors and subordinates.

Mao himself perceived the consequences of these aspects of traditional political culture early in his career and set out to replace the passivity that resulted from them with "activism." The mobilization style of leadership that has employed struggle sessions and criticism and self-criticism meetings has aimed at creating this activism, and thereby, at unleashing the creative energies of the masses from the bondage of institutional authority.[5] The emergence of highly administered complex organizations devoted to modernization in the absence of ideological growth aimed at overcoming negative traits of traditional culture are as capable of frustrating activism as were traditional patterns of authority, according to the Maoist view.

This question of activism, and of overcoming fear of authority also finds its expression in the mobilization models for science. The masses are constantly enjoined not to be overawed by the authority of bourgeois science. Such awe can only lead to passivity, and passivity in turn will both frustrate the application of science to production and will result in the emergence of a new mandarinate in the form of a technocratic elite. Instead, scientists, technologists, and science administrators must be struggled against, and their doctrines and theories must be challenged and questioned. The masses themselves must undertake research and technological innovation. In practice, this has resulted in the masses discovering the limitations of their technical competence. This realization in turn has stimulated self-initiated interest in technical education and a respect for, and desire to consult with, those with specialized training and knowledge. The self-initiation has been complemented by material and non-material incentives to pursue this behavior and an ideology that legitimizes it. Thus, science comes to the masses, but on terms that are congenial to the masses.

The conclusion that is implicit in the discussion above must now be stated explicitly. Although the policies and programs for science during the mobilization phases may have a divisive influence as a short-term effect, they appear to have the effect of expanding and integrating the science system as a longer-term consequence. This appears to be true in the functional sense that these policies help erode psychological, spatial, and organizational obstacles to communication and thereby create a more authentically systemic network of science and innovation organizations. They result in the recruitment of new participants for the research and innovative activities of the science system, and they appear to contribute on a mass scale to a sense of personal and interpersonal competence in dealing with technical authority and the risk involved in trying something new. These policies are also integrative in the sense described in the introduction, of modulating scientific and technological advance with societal and organizational change.

The Chinese Model

A question that currently concerns many students of political and economic development is "Is the Chinese model exportable or replicable?" We have attempted to describe the Chinese model for scientific development in these pages, and some tentative answers to this question can be given. It would seem that a number of elements of the Chinese model have potential applicability elsewhere. One obvious example, not unique to China, that is already receiving attention from a number of development economists is the encouragement of small-scale industry employing machinery that has been adapted to local conditions and factor endowments. Another element of the Chinese model that could be copied is simply to imitate China's level of investment in research and education, although again, this would be as much an imitation of the industrialized nations as of China. Scientific societies could be more widely encouraged in other countries to help tie together scattered and uncoordinated science-related activities with national purpose. National scientific planning could be initiated. Similarly, serious, widespread programs for the popularization of science, backed by high level support could be tried. National efforts, again backed by political purpose, to revamp the incentive and reward structures of scientists and engineers, managers, entrepreneurs, and other producers could be attempted.

However, while elements of the Chinese model could be replicated, the essence of the Chinese model is its interrelated, systemic character, the development of which we have traced to a series of policy shifts and organizational experiments. We need only look at the interrelatedness of a number of the above elements, plus others that are more difficult to replicate, to see this. The small-scale, particularly rural small-scale, industry strategy is a case in point. Apart from the more purely economic aspects of this strategy—the linkages with a highly organized agricultural sector, its reliance on local raw materials, and so forth—its reliance on adaptive technology implies a motivation and ability to adapt technology. This in turn is related to popularization programs, the existence of provincial and subprovincial R&D establishments, whose technical people interact with the production setting, local science and technology associations and experimental groups to further facilitate that interaction, and importantly, a Communist Party apparatus with a variety of means at its disposal to mobilize the relevant actors and encourage compliance.

In addition, mobilization techniques used by the Chinese, which appeal to accumulated grievances and a nationalism that is not unrelated to those grievances, may be uniquely suited to Chinese personality traits and historical experience in the sense that they touch the Chinese equivalent of the Western Protestant ethic or the Japanese reverence for emperor and nation.[6] In short, the Chinese leaders believe they have discovered an important taproot of social change. They have attempted to devise ways of linking science-related programs to it, and in the process, they have created an experience that is not readily exportable.

Problems and Prospects for the Future

There can be little doubt that in terms of the varieties of activities, the quantity of scientific information, and the levels of production within China, Chinese society has become far more complex than it was in 1949. Increasing complexity puts a premium on coordination, and coordination is typically the primary function of managerial activities.

If we reflect for a moment on our hypothetical requisites for the innovation process outlined in Chapter 1, we can see that China has at least made serious attempts at realizing most of them. One obvious area where she has apparently fallen short is in cultivating specialized managers of innovation who conform to the Maxwell's "demon" image. The functions of such managers are to express an organization's purpose in technical terms and to juxtapose that purpose against existing scientific knowledge and a technological state of the art.[7] While it is certainly true that China has been concerned with cultivating managers of innovation—witness, for instance, Ch'ien Hsueh-sen's advocacy of specialized research managers (see Chapter 4) and the Party's perennial concern for upgrading the technical knowledge of those Party cadres concerned with research and innovation—it appears that the role of innovation manager has not been formalized, and that no special recruitment and training programs have been devised.[8]

This question of management takes us to the heart of China's development problems. On one hand, the functions of the manager, to the Western mind, seem absolutely essential for development because of the coordination and integration requisites of increasing complexity in general and because of the coordination and coupling requisites in the innovative process in particular. On the other hand, because the manager sits at the nexus of the information and resource flows of multiple subsystems, his position is strategic, and considerable power can be expected to devolve to his hands in terms of the system as a whole. This devolution has been resisted in China both on political and ideological grounds. If that resistance continues, with the result that no specialized managers of innovation are produced, the crucial question then becomes: Will China soon reach an upper limit in her scientific and technological development?

The answer to this question depends on China's ability to check the inevitable trend towards structural differentiation (in terms of roles and organizations) that accompanies complexity and on her ability to devise alternative structures for performing the coordination function. If a researcher becomes a researcher-worker and either physically leaves his laboratory to spend a period of time in a factory or if the laboratory actually sets up a factory, some of the conscious coordination and coupling required in a more differentiated system will not be required. Similarly, if the socialization of researchers and workers successfully reduces the psychological distance between physical and mental labor, this too will be consistent with the checking of structural differentiation and will be a further step toward creating the kind of organic integration that will minimize the need for specialized coordination mechanisms.

Although this need can be reduced (in principle), it probably cannot be abolished. Therefore, mechanisms must be found to accomplish coordination. This is why the 3-in-1 groups and the triple combination principle in collective decision making in general bear watching. It is much too soon to predict the long-term viability and effect of this formula and indeed, as of this writing, it may be that the trend is back towards more conventional management.

To conclude, it appears that Chinese science policy has been both developmental and integrative. It may be that the integrative aspects of it have had the functional consequence of reducing the need for specialized innovation managers, a commodity in short supply in any low-income country. However, the longer-term effect of failing to produce such managers is difficult to predict.

At a different level, one suspects that Chinese efforts to achieve a high degree of integration of science, technology, production, societal change, ideology, and even morality will continue to have developmental consequences for the remainder of the twentieth century. As we in the industrialized countries approach the twenty-first century, we sense a growing crisis in science-society relations. Scientific and technological advances have outpaced the capabilities of our institutions. Will China's continuing scientific development lead to the same kind of crisis of integration now evident in the industrialized nations? Or, will the industrialized nations begin to look to China for guidance in resolving the science-society problems of the late twentieth and early twenty-first centuries? Our present inability to provide answers in no way detracts from the importance of the questions.

Notes

1. The exact number of institutions doing R&D work is difficult to establish. According to a 1958 report, there were 840 research institutes, See Chu-yüan Cheng, *Scientific and Engineering Manpower in Communist China, 1949-1963* (Washington: National Science Foundation, 1965), p. 26. Since the 840 figure included institutions under CAS branch academies, and since many of those were either abolished or consolidated after the Great Leap Forward, the actual number is probably somewhat less. In Ralph J. Watkins (ed.), *A Directory of Selected Scientific Institutions in Mainland China* (Stanford: The Hoover Institution Press, 1971), 490 institutions are identified. That figure is ambiguous, however, because it includes professional societies that are not strictly research performers. Furthermore, as the title of the directory suggests, the 490 figure represents a selected sample. It seems safe to say that the actual number is well over 500, which supports the point being made here that China's progress in this regard has been impressive.

2. Leo A. Orleans, "Research and Development in Communist China: Mood, Management, and Measurement," in U.S. Joint Economic Committee, *An Economic Profile of Mainland China* (New York: Frederick A. Praeger, Publishers, 1968), p. 577.

3. Yuan-li Wu and Robert B. Sheeks, *The Organization and Support of Scientific Research and Development in Mainland China* (New York: Praeger Publishers, 1970), pp. 198 and 224.

4. Chu-yüan Cheng. "Scientific and Engineering Manpower in Communist China," in U.S. Joint Economic Committee, op. cit., p. 529.

5. This political culture interpretation of political development is drawn from the findings of Richard Solomon as reported in "Communications Patterns and the Chinese Revolution," *The China Quarterly* 32 (October-December, 1967); "On Activism and Activists," 39 (July-September, 1969); and *Mao's Revolution and the Chinese Political Culture* (Berkeley: University of California Press, 1971). Although Solomon is not explicitly concerned with administration, the implications of his work for administration and organizational behavior are significant.

6. Cf. Solomon, "On Activism and Activists," p. 107.

7. Cf. J.A. Morton, *Organizing for Innovation* (New York: McGraw Hill Book Company, 1971), p. 46ff.

8. Richman has noted a serious shortage of specialized managerial talent in the industrial sector generally. See Barry M. Richman, *Industrial Society in Communist China* (New York: Vintage Books, 1969), p. 819ff.

Bibliography

Bibliography

Primary Sources

Frequently occurring titles have been abbreviated according to the standard usage in works on contemporary China. These include:

JMJP. *Jen-min Jih-pao (People's Daily).*
KMJP. *Kuang-ming Jih-pao (Kuangming Daily).*
NCNA. *New China News Agency.*
KHTP. *K'o-hsueh T'ung-pao (Science Bulletin).*

Standard translation services used include:

1. U.S. Consulate General, Hong Kong.
   SCMP. Survey of the China Mainland Press.
   CB. Current Background.
   ECMM. Extracts from China Mainland Magazines, superseded by
   SCMM. Selections from China Mainland Magazines.
2. U.S. Department of Commerce.
   JPRS. Joint Publications Research Service.

"Academic Activities in China." *KHTP* 3 (March 17, 1962). In *SCMM* 313.

"Academic Committee Set Up by Petroleum Research Institute of the Chinese Academy of Sciences." *JMJP.* October 20, 1955. In *SCMP* 1162.

"Academy of Sciences Announces Five-Year Plan." *NCNA*, July 24, 1955. In *SCMP* 1068.

"Academy of Sciences Convenes Forum for Peking Scientists to Discuss Contradictions Within Their Ranks." *NCNA*. May 3, 1957. In *SCMP* 1541.

"Academy of Science Entertaining Soviet Experts." *NCNA.* November 7, 1957. In *SCMP* 1651.

"Actively Unfold Scientific Study and Discussion of Technological Policies." *Hung Ch'i* 11 (June 9, 1962). In *SCMM* 319.

"Agro-Scientific Research Institute in Sungari Area Vigorously Pushes Rural Mass Activities of Scientific Experiments." *JMJP.* January 31, 1973. In *SCMP* 5315.

"Announcements of Chinese Academy of Sciences on Enrollment of Post-Graduates in 1957." *KMJP.* October 30, 1957. In *SCMP* 1637.

"Announcement of Research Institute of Electronics on Enrollment of Post-Graduates." *KMJP.* October 12, 1957. In *SCMP* 1640.

"Association of Science and Technology of Shanghai Sums Up and Popularizes Achievements in Technical Innovation." *JMJP.* September 8, 1960. In *SCMP* 2340.

"Background Material: Solar Eclipses and Their Research." *NCNA*. January 23, 1969. In *SCMP* 4348.

"The Beginnings of a New Stage of Our Scientific Undertaking." *KMJP* (editorial). June 3, 1955. In *SCMP* 1068.

"Build a Powerful Force of Scientific and Technical Workers." *JMJP* (editorial). May 29, 1964. In *SCMP* 3236.

Chang P'eng. "The Situation on the Discussion of the Question of the Criterion of Practice." *Hung Ch'i* (May 23, 1964). In *SCMM* 422.

Chang Chi-tao. "How Wang Pao-ching Made His Farm a Laboratory." *China Reconstructs* (October 1960), pp. 32-34.

Chang Chien. "Strive to Train a Large Army of 'Red and Expert' Scientists and Technologists." *Chi-hua Ching-chi (Planned Economy).* May 9, 1958. In *ECMM* 135.

Chang Ching-fu. "For the Liberation of Thought and the Determined Follow Through of the Main Party Line Concerning the Construction of Socialism in the Tasks of Science." *JMJP.* June 7, 1958. In *JPRS* DC 472.

Chang Kuang-tou. "The Characteristics and Tasks of Technical Science." *JMJP.* August 9, 1962. In *JPRS* 15569.

Chang Wan-chang and Lin Shih-kao. "The Method for Conducting Scientific Experiments in Conjunction with Production." *Chung-kuo Nung-yeh k'o-hsueh (Chinese Agricultural Science).* (March 15, 1965). In *SCMM* 468.

Chao Shih-ying. "Some New Tendencies in the Development of Sciences." *Hsin Chien-she (New Construction)* (February 20, 1966). In *SCMM* 545.

Ch'en Feng-chi. "Scientific and Technical Education in China." *Ajiya Kenkyu (Asia Research)* 387 (Tokyo, 10 September 1963). In *JPRS* 22, 237.

Ch'en Hua-k'uei. "Agricultural Production and Agricultural Science."*Hung Ch'i* (April 16, 1963). In *SCMM* 364.

Ch'en Kuo-ta. "How to Stringently Organize the Process of Scientific Research." *KMJP.* November 1963. In *JPRS* 22418.

Ch'en Po-ta. *Speech Before the Study Group of Research Members of Academia Sinica.* Peking: Foreign Language Press, 1952.

Ch'en Shih-ho. "Shanghai's Technical Revolution Scores New Victories." *China Reconstructs* 13 (May 1965).

Ch'eng Li-chia. "Chemical Research Institute of Chinese Academy of Sciences Makes a Good Job of Administrative and Auxiliary Work in Coordination With the Needs of Research Work." *KMJP.* September 20, 1961. In *SCMP* 2598.

Chiang Chiu-chin. "Science and Technology Association of Shanghai Spreads Scientific and Technical Knowledge of Agriculture." *KMJP.* August 25, 1962. In *SCMP* 2823.

Ch'ien Hsueh-sen. "Large-Scale Scientific Experimental Work." *JMJP.* August 30, 1964. In *SCMP.* 3297.

_______. "Organization and Management Work in Science and Technology."

Hung Ch'i 22 (November 19, 1963). in *JPRS* 22660.

Ch'ien San-ch'iang. "The Development of Natural Science in New China." *People's China* (September 16, 1953). In *SCMP* 673.

Chin Shan-pao. "Demonstration Fields–Bases for Agricultural Research." *China Reconstructs* (June 1965).

"China Grants Science Awards for the First Time." *NCNA*. January 24, 1957. In *CB* 437.

"China's Natural Sciences Circles Very Active." *Chung-kuo Hsin-wen* (China News). February 8, 1965. In *JPRS* 29359.

"China's Success in Making Large Scale Comprehensive Observation of a Total Eclipse." *NCNA*. January 23, 1969. In *SCMP* 4348.

"China's Unique Road for Developing Science and Technology." *NCNA*. October 17, 1966. In *SCMP* 3805.

"Chinese Academy of Sciences Calls Conference of Institute Directors: Key-Point Research Subjects Decided On." *NCNA*. November 22, 1953. In *SCMP* 695.

"Chinese Academy of Sciences Convenes Meeting of Research Institute Directors to Plan Big Jump in Scientific Work." *NCNA*. February 20, 1958. In *SCMP* 1728.

"Chinese Academy of Sciences Grants Awards to 19 Research Units and 23 Individuals." *NCNA*. June 4, 1955. In *SCMP* 1062.

"Chinese Academy of Sciences Held Conference for Exchange of Work Experience." *JMJP*. May 29, 1964. In *SCMP* 3236.

"Chinese Academy of Sciences Sets up Secretariat to Strengthen Leadership of Scientific Research Work." *NCNA*. April 14, 1954. In *SCMP* 792.

"Chinese Academy of Sciences Tables Draft Five-Year Plan." *NCNA*. June 3, 1955. In *SCMP* 1062.

"Chinese Academy of Sciences to Set Up Four Departments to Guide Scientific Research." *China News Service*. April 13, 1954. In *SCMP* 788.

"Chinese Research to be Decentralized." *NCNA*. June 13, 1957. In *SCMP* 1554.

Chou Kuang-chao and Ho Tso-hsiu. "Theory and Experiment in the Study of Physics." *Hung Ch'i* (May 20, 1963). In *SCMM* 369.

Chou Kuang-yu. "The Function of a Tutor in Communist China." *KHTP*. April 11, 1960. In *JPRS* 6058.

Chou Pei-yuan. "China's Progress in Physics." *K'o hsueh Hsin-wen (Science News)* (28 September 1959). In *JPRS* 3116.

Chu K'o-chen. *Culture and Education in New China.* Peking: Foreign Language Press, 1950. In *CB* 153.

Chu K'o-chen. "Scientific Research and University Education." *Ta Kung Pao* (Shanghai). October 29, 1950. In *CB* 153.

Chu Te-hai. "Extensively Unfold Mass Activities in Agricultural Scientific Experiments and Promote a New Upsurge in Agricultural Production." *JMJP*. January 31, 1965. In *SCMP* 3396.

"Combination of Activities Both Inside and Outside the Institute." *JMJP.* August 25, 1972.

"Conclusion of Enlarged Meeting of the Scientific Planning Commission of the State Council." *NCNA.* June 15, 1957. In *SCMP* 1564.

"Conduct Research Realistically and Assiduously; Study Without Being Bored, Teach Without Being Wearied." *KMJP.* January 28, 1962. In *SCMP* 2684.

"The Conference on Exchange of Supplies for Scientific Research." *KMJP* (editorial). November 4, 1963. In *SCMP* 3114.

"Congratulations to the Chinese Academy of Sciences on the Inauguration of Departments." *KHTP* (editorial). 6 (1955). In *SCMP* 1092.

"Correct Way on Conducting Agricultural Scientific Experiments." *JMJP.* July 23, 1963. In *SCMP* 3037.

"A Criticism of the Theory of Making the Electronics Industry the Center." *JMJP.* August 12, 1971.

"Criticisms and Suggestions made by Scientists of Department Meetings of Department Committee of the Academy of Sciences." *JMJP.* May 27, 1957. In *CB* 460.

"Cultural and Scientific and Technical Units of Peking Municipality Hold Various Classes to Help the Cadres and the Masses Raise Their Theoretical Level and Work Ability." *KMJP.* February 11, 1973. In *SCMP* 5319.

"Decision Made at First Provincial Scientific Conference of Kiangsi." *KMJP.* August 2, 1958. In *SCMP* 1842.

"Deliberate Castration and Shameless Betrayal–A Criticism of Liu Shao-ch'i's and Other Political Swindler's Theory of 'Electronics as the Core'." *KMJP.* December 13, 1971. In *SCMP* 5045.

"Directive on the Drawing Up of a 15-Year Long-Range Plan of the Chinese Academy of Sciences." *KHTP* 11 (1955). In *ECMM* 20.

"Doggedly Combine Theory with Practice and Perseveringly Follow the 'May 7' Road." *JMJP.* February 17, 1973. In *SCMP* 5323.

Du Ruen-Sheng. (Tu Jun-sheng) "Great Progress Made in the Natural Sciences in China During the Last Decade." *Scientia Sinica* VIII, 11 (November 1959), pp. 1195-1217.

"Economists Plan This Year's Research." *NCNA.* March 26, 1958. In *SCMP* 1745.

"Encourage Inventions and Technical Improvements by Rewards in Order to Promote Development of our Production and Construction." *JMJP.* December 2, 1963. In *SCMP* 3117.

"Enlarged Meeting of National Committee of All-China Federation of Scientific Societies Concludes." *NCNA.* July 20, 1957. In *SCMP* 1587.

"Explanation of the Procedure Governing the Adjustment of the 1956 Science Awards (Natural Science Section) of Chinese Academy of Sciences." *JMJP.* January 25, 1957. In *CB* 437.

Fan Ch'ang-chiang. "The Question of 'Motive Power' of the Intellectuals." *Chung-kuo Ch'ing-nien (Chinese Youth)* (June 1, 1958). In *ECMM* 142.

Fang Fu-lin. "What I've Learned in Eleven Years." *China Reconstructs* (June 1963), pp. 24-26.

Fang Yu-ting. "We are the Masters of Our Plant." *China Reconstructs* 8 (March 1962).

"Fifteen Years of Scientific Development Under Mao Tse-tung's Leadership." *KHTP*. October 1964. In *JPRS* 28136.

"First National Conference on Activists in the Dissemination of Scientific and Technical Knowledge Opens." *NCNA* (October 29, 1956). In *SCMP* 1412.

"First Science Awards of China." *JMJP* (editorial). January 25, 1957. In *CB* 437.

"General Resolution Adopted at Meeting of Chinese Academy of Sciences, 15-Year Plan Called For." *NCNA*. June 11, 1955. In *SCMP* 1068.

"Geophysics Conference Closes." *NCNA*. February 9, 1957. In *SCMP* 1469.

"Great Achievements in Spread of Chinese Science." *Chung-kuo Hsin-wen*. October 5, 1965. In *JPRS* 32912.

"Hail Warmly the Splendid Gains in the Revolution in Education and Implement More Firmly Chairman Mao's Proletarian Line for Education." *JMJP*. January 5, 1973. In *SCMP* 5296.

"Higher [Education] Institutions and Scientific Research Organs to Further Cooperate in Scientific Research Work." *NCNA*. January 15, 1956. In *SCMP* 1217.

Ho Tsu-hsiu. "Natural Science and Practical Application." *Hung Ch'i* 7 (April 1, 1962). In *SCMM* 311.

Ho Tsu-hsiu. "The Roles of Experiments, Abstractions, and Hypotheses in Scientific Research." *Hung Ch'i* 11 (June 1, 1961). In *SCMM* 266.

Hou Te-pang. "Address Delivered at Opening Ceremony of the Polyvinyl Chloride Meeting of China Chemistry Society and China Chemical Engineering Society." *Hua-hsueh T'ung-pao (Chemistry Bulletin)* (December 1965). In *JPRS* 34696.

Hsing Pen-ssu. "Collective Research and Individual Research." *Hung Ch'i* (October 1, 1962). In *SCMM* 337.

Hsu Hsing. "Academia Sinica Holds Shanghai Conference on Submerged Paddy Research Reports." *Chih-wu Shengli-hsueh T'ung-hsun* 4 (August 1959). In *JPRS* 2388.

Hsu Nai-an. "The First All-China Conference on Hydrographic Geology and Engineering Geology." *KHTP*. April 4, 1965.

Hua Lo-keng. "Notes on Experimental Work in the Method of Overall Planning." *Hung Ch'i* (October 1, 1965). In *SCMM* 496.

Huang Ming-lung. "Step Up Research Work and Raise the Academic Standard." *KMJP*. March 30, 1962. In *SCMP* 2723.

Hung Chi-kung. "The Working Class Must Firmly Grasp Power of Leadership Over Science and Technology." *KMJP*. December 16, 1968. In *SCMP* 4332.

"Important Steps to Strengthen Leadership Over Scientific Work." *JMJP* (editorial). June 3, 1955. In *SCMP* 1064.

"Institute of Physics of the Academy of Sciences Formulates Regulations and Systems to Guarantee the Research Personnel Time for Attending to Their Research." *KMJP.* September 20, 1961. In *SCMP* 2598.

"Is it Good for the Working Class to Have Control of Science and Technology." *KMJP.* March 21, 1969. In *SCMP* 4392.

Jen Wen. "Road of Training Technicians From Among the Workers." *Peking Review* 32 (August 6, 1971), pp. 12-15.

Kao Shang-yin. "Microbiology in China, and in Particular, Virological Research at the Institute for Microbiology in Wuhan." *Priroda* (Nature) (December 1962). In *JPRS* 19523.

Kuo Mo-jo. "Anti-Rightist Struggle of Social Scientists Must be Further Intensified." *JMJP.* September 19, 1957. In *SCMP* 1634.

_______. "Closing Address." (Given at Second Plenum of Department of Committee). *JMJP.* May 31, 1957. In *CB* 460.

_______. "Development of Scientific Research in China." *NCNA.* June 18, 1956. In *CB* 400.

_______. "General Report on 1950 Work and Main Points of 1951 Program of the China Academy of Sciences." (Approved by the 70th Meeting of the Government Administrative Council, February 2, 1951). *Hsin-hua Yueh-pao (New China Monthly)* (May 1951). In *CB* 153.

_______. "In Refutation of An Anti-Socialist Scientific Program." *JMJP.* July 6, 1957. In *CB* 467.

_______. "On Coordination of Scientific Research." *JMJP.* March 20, 1957. In *SCMP* 1515.

_______. "Report Concerning the Basic Conditions of the Chinese Academy of Sciences and Its Future Work." (Given at the 204th Meeting of Government Administrative Council, January 28, 1954). *JMJP.* March 26, 1954. In *CB* 359.

_______. "Report Delivered at Inaugural Meeting of Departments of Chinese Academy of Sciences." *JMJP.* June 2, 1955. In *CB* 359.

"Kuo Mo-jo Encourages Contention Among Scientists." *NCNA.* July 17, 1957. In *SCMP* 1574.

"Kuo Mo-jo Speech to National People's Congress, July 21, 1955." *NCNA.* July 21, 1955. In *CB* 351.

Kung Yü-chih. "On Scientific Experiment." *Hung Ch'i* 1 (January 6, 1965). In *SCMM* 454.

"Lay a Foundation, Labor Hard, and Develop the Motherland's Science and Technology." *Chung-kuo Ch'ing-nien Pao (China Youth News)* (January 4, 1962). In *SCMP* 2663.

Li Fu-chun. "On Importance of Popularizing Science." *NCNA.* October 30, 1950. In *SCMP* 1403.

Li Hsin-kung. "Technical Power is in the Firm Grip of Us Workers." *Hung Ch'i* (September 10, 1968). In *SCMM* 629.

Li Ssu-Kuang. "Developments in Geological Work in Communist China in the Past Decade." *KHTP* 18 (1956). In *JPRS* 5090.

Li Tsai-ming. "How Forward Leaps were Accomplished at the Laboratory of Semiconductors." *KHTP* 18 (September 27, 1958). In *JPRS* 1479.

Liang Ssu-ch'eng. "Which Should Come First—'Becoming Red' or 'Becoming an Expert'." *Chung-kuo Ching-nien (Chinese Youth)* (December 1, 1957). In *ECMM* 121.

"The Library System of Chinese Academy of Sciences." *T'u-shu-Kuan (Library)* (September 1964). In *JPRS* 28136.

Liu Hsien-chou. "A Few Questions in Scientific Research Work." *JMJP.* December 19, 1957. In *SCMP* 1703.

Liu Keng-ling and Chiang Chao-yü. "Research in Agricultural Sciences Must be Combined with Production Practice." *Che-hsüeh Yen-chiu (Philosophical Studies)* (March 16, 1965). In *SCMM* 474.

Lo Ying. "Achieve Rejuvenation by Our Own Efforts, Make Haste to Catch Up with the Advanced." *KMJP.* April 30, 1963. In *SCMP* 2983.

Lu Hsin-ch'u. "The Exuberant Atmosphere of the Science Hall." *JMJP.* March 25, 1963. In *JPRS* 19523.

Lu Hsueh-shan. "Deputy Director Academia Sinica Physics Institute Discusses—Theories, Practices, Interests and Others." *KMJP.* December 22, 1962. In *JPRS* 18023.

Lu K'uei-hung. "On Absoluteness and Relativity of Criteria for Practice—About a Chemical Factory's Three Kinds of Test Fields." *Hung ch'i* (June 30, 1964). In *SCMM* 428.

Lu Ming. "Scientific Experiments and Development of Modern Sciences." *Chung-kuo Ch'ing-nien (Chinese Youth)* (May 16, 1964). In *SCMM* 423.

Ma Hua. "National Conference on Field Practice Work in Agricultural Sciences." *KHTP.* May 1965. In *JPRS* 31564.

Mao I-sheng. "Knowledge and Experience of Scientific Experiment." *Hsin Chien-she (New Construction)* 11-12 (December 20, 1965).

"Mao Tse-tung Illuminates the Road of Achieving Greater, Faster, Better and More Economical Results in Geological Work." *JMJP.* February 14, 1971. In *SCMP* 4846.

Mao Tse-tung. *The Selected Works of Mao Tse-tung.* Peking: Foreign Language Press, 1965.

"Meeting of USSR Academy of Sciences and Council of Scientific-Technical Societies." *Nauchno-tekhnicheskiye obshchestvy SSSR* Scientific-Technical Societies of the USSR). In *JPRS* 44562.

" 'Miners' Scientific Research Institute Achieves Important Successes." *NCNA.* April 6, 1971. In *SCMP* 4878.

"More Libraries to Help Scientists." *NCNA.* June 15, 1957. In *SCMP* 1553.

"National Conference Exchanges Rural Scientific Experience." *KMJP.* May 8, 1966. In *JPRS* 35755.

"National Science and Technology Association Holds Meeting in Peking on Mass Experimental Activities." *NCNA*. May 20, 1964. In *SCMP* 3231.

"New Achievements of Our Modernized Laboratories" *KMJP*. November 10, 1963. In *JPRS* 22418.

"New Scientific Institutes to Be Set Up." *NCNA*. November 1, 1956. In *SCMP* 1405.

Nieh Jung-chen. "Let the Whole Party Lay a Firm Hold on Scientific and Technological Work to Carry Out the Technological Revolution." *JMJP*. August 2, 1958. In *SCMP* 1831.

_______. "The Realization of a Big Leap Forward in Science." *NCNA*. March 14, 1958. In *SCMP* 1745.

"A Number of Scientific Advanced Units and Individuals Receive Awards." *JMJP*. May 29, 1964. In *JPRS* 25263.

"Numerous Contradictions Within the Rank of the Scientific Circle." *NCNA*. April 30, 1957. In *SCMP* 1541.

"Party Committee of Peking University Organizes the Teachers, Students and Employees to Read Books and Learn Seriously, Repudiates Revisionism and Rectifies Working Style Penetratingly and Pushes Forward the Development of Educational Revolution." *JMJP*. January 12, 1973. In *SCMP* 5302.

P'ei Yu. "Establish a Good Link-Up Between Factory and School in an Earnest Way." *Hung Ch'i* 9 (September 1, 1972). In *SCMM* 737-38.

"People Encouraged to Call on Government Departments." *NCNA*. May 11, 1957. In *SCMP* 1530.

"People Use Mao Tse-tung Thought to Study Solar Eclipse." *NCNA*. January 24, 1969. In *SCMP* 4349.

"Positively Cultivate the New Source of Strength in Scientific Research Work." *JMJP* (editorial). September 6, 1955. In *SCMP* 1129.

"Professors of Peking University Start Scientific Research Work." *KMJP*. May 8, 1954. In *SCMP* 821.

"Raise the Level of the Academic Activities of Natural Science Societies." *KMJP* (editorial). September 9, 1962. In *SCMP* 2832.

"Regulations on Awards for Inventions." *JMJP*. December 2, 1963. In *SCMP* 3117.

"Regulations on Awards to Scientists." *NCNA*. September 1, 1955. In *SCMP* 1122.

"Regulations on the Organization of the Departments of the Chinese Academy of Sciences." In Cheng Chu-yüan. *Scientific and Engineering, Manpower in Communist China, 1949-1963*. Washington: National Science Foundation, 1965, pp. 302-304.

"Report of Kuo-Mo-jo at Inaugural Meeting of Departments of Chinese Academy of Sciences." *NCNA*, June 2, 1955. In *SCMP* 1066.

"Resolution Concerning Re-inforcing the Editorial Work of *K'o-hsueh T'ung-pao*". *KHTP*. September 19, 1963. In *JPRS* 23592.

"Resolution on the Establishment of the Science and Technology Association." In U.S. Department of Commerce. *JPRS* 1502 DL.

"Revival Through Self-Exertion and Acceleration of Construction of Modernized Laboratories." *KMJP* (editorial). November 10, 1963. In *JPRS* 22418.

"The Revolution in Education in Colleges of Science and Engineering as Reflected in the Struggle Between the Two Lines at the Shanghai Institute of Mechanical Engineering. *Hung Ch'i* (September 1968). In *SCMM* 627.

Revolutionary Committee of Hsiyang *Hsien*, Shansi. "Persevere in Scientific Experiment and Develop Agricultural Production." *JMJP*. December 29, 1972. In *SCMP* 5293.

Revolutionary Mass Criticism Group of Kirin University. "Deliberate Castration and Shameless Betrayal." *KMJP*. December 13, 1971. In *SCMP* 5045.

"The Road for the Training of Scientific and Technological Personnel in Rural Areas." *JMJP*. March 14, 1969. In *SCMP* 4389.

"Science Should Help Giant Leap Forward." *NCNA*. March 12, 1958. In *SCMP* 1735.

"Scientific Conference Begins Open Discussion." *NCNA*. May 27, 1957. In *SCMP* 1541.

"Scientific Planning Committee Holds Meeting." *NCNA*. June 13, 1957. In *SCMP* 1564.

"Scientific Research to Help Industry." *NCNA*. June 9, 1955. In *SCMP* 1068.

"Scientists Continue to Hit Back to Rightists." *NCNA*. July 16, 1957. In *SCMP* 1581.

"Scientists Denounce Rightists at Meeting." *NCNA*. July 16, 1957. In *SCMP* 1594.

"Second Group of Postgraduate Students Enrolled by CAS." *NCNA*. April 26, 1957. In *SCMP* 1542.

"Second Plenum of Academic Department Committee of Chinese Academy of Science Opens May 23," *JMJP*. May 24, 1957. In *CB* 460.

"Seriously Study Chairman Mao's Latest Instruction, Pledge to Carry the Proletarian Revolution in Education Through to the End." *Pei-ching Jih-pao (Peking Daily)*. July 26, 1968. In *SCMM* 654.

"Seven Research Institutes of the Chinese Academy of Sciences Set Up Academic Committees." *NCNA*. December 26, 1955. In *SCMP* 1201.

"Shanghai Scientists Discuss 12 Year Plan." *Wen Hui Pao*. March 9, 1957. In *SCMP* 1500.

"Shen Hsien-lun. "Cooperation in Scientific Research." *KMJP*. April 24, 1964. In *JPRS* 25070.

Shih Yen-hung. "Down with the Chief Backer of the Revisionist Educational Line." *JMJP*. July 18, 1967. In *CB* 836.

"Speech at the Inaugural Meeting of the Natural Science Research Society of the Border Region." *NCNA*. January 19, 1967. In *SCMP* 3866.

"State Council Decision of 16 June 1956 Regarding Wage Reform." *Kung-jen Jih-pao (Worker Daily)*. 5 July 1956. In *JPRS* 585 N.Y.

"State Council's Scientific Planning Committee Holds Third Enlarged Meeting." *NCNA*. December 29, 1956. In *SCMP* 1453.

"Steadily Develop Mass Scientific Experiment Activities in the Countryside." *JMJP* editorial. May 21, 1964. In *SCMP* 3231.

"Strengthen the Work of the Popularization of Scientific Techniques to Ensure the Fulfillment of the First Five-Year Plan." *KMJP* (August 26, 1955). In *SCMP* 1129.

Sun Ming-hui. "New Horizons for Innovators." *China Reconstructs* (May 1963), pp. 2-4.

Sun Shu-jen. "The Danger of Bourgeois Fame and Wealth Thinking to Scientific Undertakings." *KMJP*. October 15, 1964. In *SCMP* 3334.

Tai Sung-en. "Directions of Agricultural Research." *China Reconstructs* (April 1964), pp. 14-17.

T'ang Ao-ch'ing. "The Thought of Mao Tse-tung Is a Sharp Weapon in Studying Natural Science." *Hung Ch'i* (October 1, 1965). In *SCMM* 496.

T'ao Meng-ho, "China's New Academia Sinica." *China Monthly Review* (November 1951). In *CB* 153.

"There are Good Prospects for Factories to Undertake Scientific Research." *KMJP*. February 28, 1971. In *SCMP* 4857.

"There is Nothing Mysterious About Science." *JMJP* (editorial). May 22, 1958.

Ti'en Fu. "Ideological Emancipation and Scientific Progress." *Hsueh Hsi (Study)* (July 3, 1958). In *ECMM* 144.

"To Lead Mechanization with Revolutionization is the Broad Level Road to the Development of the Machine Building Industry." *KMJP*. December 16, 1968. In *SCMP* 4332.

"Vice Premier Nieh Jung-chen Gives Important Directions on Agricultural Science." *NCNA*. November 28, 1957. In *SCMP* 1670.

Wang Chia-ch'ien and Wang Chiung-yeh. "What Chemistry Can Contribute to Agriculture." *K'o-hsueh Ta-chung* (May 5, 1963). In *SCMM* 369.

"Wang Lin-hao. "Take Our Own Road and Scale the Pinnacle of World Science and Technology." *Tien Shih Chieh* (August 15, 1966). In *SCMM* 12.

Wu Chung-wei. "The Procedure Problem in Scientific Research Work." *KMJP*. June 4, 1964. In *JPRS* 25, 337.

Yang Ch'ao-shu. "Intermediate Tests in Science and Technology." *Hung Ch'i* (April 20, 1964). In *SCMM* 417.

_______. "On the Problem of Selection and Training of Scientific Personnel." *Hung Ch'i* (September 1, 1962). In *SCMM* 332.

Yang Wen. "Second and Third Enlarged Council Meetings of the Meteorological Society of China." *KHTP*. August 27, 1958. In *JPRS* 772 NL.

"A Year of New Achievements and Many Successes in Research for Scientists in China." *NCNA*. January 13, 1963. In *SCMP* 2906.

Yen Chung-p'ing. "On the Selection of Subjects of Research." *Hung Ch'i* (June 16, 1962). In *SCMM* 321.

Yu Ch'i-pao. "Some Points Concerning Research to be Carried Out Jointly by Agronomists and the Masses." *Hung Ch'i* 22 (November 19, 1963). In *JPRS* 22660.

Yu Kuang-yuan. "On Mass Movements in Scientific and Technological Work." *Hung Ch'i* 3 (February 1, 1960). In *JPRS* 3169.

_______. "Two Roads for Developing the Scientific Enterprise in China." *JMJP.* September 4, 1957. In *SCMP* 1620.

Secondary Sources

Andors, Stephen. "Revolution and Modernization: Man and Machine in Industrializing Society, The Chinese Case." In Edward Friedman and Mark Selden (eds.). *America's Asia.* New York: Vintage, 1971.

Aran, Lydia and Joseph Ben-David. "Socialization and Career Patterns as Determinants of Productivity of Medical Researchers." *Journal of Health and Social Behavior* 9 (March 1968), pp. 3-15.

Argyris, Chris. *Organization and Innovation.* Homewood, Illinois: The Dorsey Press, 1965.

Barnett, A. Doak. *Cadres, Bureaucracy and Political Power in Communist China.* New York: Columbia University Press, 1967.

_______. "Mechanisms for Party Control in the Government Bureaucracy in China." In Fred W. Riggs (ed.). *Frontiers of Development Administration.* Durham, North Carolina: Duke University Press, 1970.

Berliner, Joseph S. "Bureaucratic Conservatism and Creativity in the Soviet Economy." In Fred W. Riggs (ed.), *Frontiers of Development Administration.* Durham: Duke University Press, 1970.

Board on Science and Technology for International Development, Office of the Foreign Secretary. *Research Management and Technical Entrepreneurship.* Washington: National Academy of Sciences, 1973.

Boorman, Howard L. (ed.). *Biographical Dictionary of Republican China.* Vol. I (Ai-Ch'i). New York: Columbia University Press, 1967.

Center for International Affairs and the East Asian Research Center. *Communist China 1955-1959: Policy Documents with Analysis.* Cambridge, Massachusetts: Harvard University Press, 1965.

Chakrabarti, Alok K. *The Effects of Techno-economic and Organizational Factors on Adoption of NASA Innovations by Commercial Firms in the United States.* Unpublished Ph.D. dissertation. Northwestern University, 1972.

Chang Kuei-sheng. "Nuclei-formation on Communist China's Iron and Steel Industry." *Annals of the Association of American Geographers* LX, 2 (June 1970), pp. 257-285.

Chao Kang. "Policies and Performance in Industry." In Alexander Eckstein, Walter Galenson, and Ta-Chung Liu (eds.). *Economic Trends in Communist*

China. Chicago: Aldine Publishing Company, 1968.

Chen, C.S. (ed.). *Rural People's Communes in Lien-Chiang.* (trans. Charles Price Ridley). Stanford: Hoover Institution Press, 1969.

Chen Pi-chao. "Over-urbanization, Rustication of Urban-Educated Youths, and the Politics of Rural Transformation: The Case of China." *Comparative Politics* IV, 3 (April 1972), pp. 361-386.

Chen, Theodore H.E. *Thought Reform of the Chinese Intellectuals.* Hong Kong: Hong Kong University Press, 1950.

Cheng Chu-yüan. *The Machine-Building Industry in Communist China.* Chicago: Aldine-Atherton, 1971.

_______. *Scientific and Engineering Manpower in Communist China, 1949-1963.* Washington, D.C.: National Science Foundation, 1965.

_______. "Scientific and Engineering Manpower in Communist China." In Joint Economic Committee of the U.S. Congress. *An Economic Profile of Mainland China.* New York: Frederick A. Praeger, 1967.

Christiansen, Wilbur N. "Science and Scientists in China Today." *Scientific Research* (October 1967), pp. 64-68.

Chuka Jinmin Kyowakoku Soshiki Betsu Jinmeihyo (Tables of Personnel and Government Organizations of the Chinese People's Republic). Tokyo: Naikaku Kambo Kaikaku Chosashitsu, 1957, 1959, 1961.

Cooper, Charles. "Science, Technology and Production in the Underdeveloped Countries: an Introduction." *The Journal of Development Studies* XIX, 1 (October 1972), pp. 1-18.

Croizier, Ralph C. *Traditional Medicine in Modern China.* Cambridge: Harvard University Press, 1968.

Crook, Frederick W. "Chinese Communist Agricultural Incentive Systems and the Labor Productive Contracts to Households: 1956-1965." *Asian Survey* XIII, 5 (May 1973), pp. 470-481.

Dailey, Charles A. *Entrepreneurial Management.* New York: McGraw-Hill, 1971.

Dean, Genevieve. "A Note on the Sources of Technical Innovations in the People's Republic of China." *The Journal of Development Studies* IX, 1 (October 1972), pp. 187-198.

_______. "China's Technological Development." *New Scientist* (May 18, 1972), pp. 371-373.

_______. "Science, Technology and Development: China as a 'Case Study'." *The China Quarterly* 51 (July-September 1972), pp. 520-534.

Dedijer, Steven. "Underdeveloped Science in Underdeveloped Countries." *Minerva* II, i (Autumn 1963), pp. 61-81.

Dernberger, Robert F. "Radical Ideology and Economic Development in China: The Cultural Revolution and its Impact on the Economy." *Asian Survey* XII, 12 (December 1972), pp. 1048-1065.

Deutsch, Karl W. "Social Mobilization and Political Development." *American Political Science Review* LV (September 1961).

De Witt, Nicholas. "Reorganization of Science and Research in the U.S.S.R."

Science 133, 3469 (June 23, 1961), pp. 1981-1987.

_______. "Soviet Science: The Institutional Debate." *Bulletin of the Atomic Scientists* (June 1960), pp. 208-211.

Donnithorne, Audrey. "China's Cellular Economy: Some Economic Trends Since the Cultural Revolution." *The China Quarterly* 52 (October-December 1972), pp. 605-619.

_______. *China's Economic System.* New York: Praeger, 1967.

Eckstein, Alexander. "Economic Fluctuations in Communist China's Domestic Development." In Ping-ti Ho and T'ang Tsou (eds.). *China in Crisis.* Vol. 1, Book 2. Chicago: University of Chicago Press, 1968.

Eckstein, Alexander, Walter Galenson, and Ta-chung Liu (eds.). *Economic Trends in Communist China.* Chicago: Aldine, 1968.

Ellul, Jacques. *The Technological Society* (trans. John Wilkinson). New York: Alfred A. Knopf, 1964.

Etō, Shinkichi. "Communist China: Moderation and Reform in the Chinese Revolution." In James B. Crowley (ed.). *Modern East Asia: Essays in Interpretation.* New York: Harcourt, Brace and World, Inc., 1970.

Folger, Anne and Gerald Gordon. "Scientific Accomplishment and Social Organization." *American Behavioral Scientist* 6 (December 1962), pp. 51-58.

Graham, Loren R. "Reorganization of the U.S.S.R. Academy of Sciences." In Peter H. Jamiler and Henry W. Norton (eds.). *Soviet Policy-making.* New York: Praeger Publishers, 1967.

_______. *The Soviet Academy of Sciences and the Communist Party, 1927-1932.* Princeton: Princeton University Press, 1967.

Greenblatt, Sidney Leonard. "Organizational Elites and Social Change at Peking University." In Robert A. Scalapino (ed.). *Elites in the People's Republic of China.* Seattle: University of Washington Press, 1972.

Gruber, William H. and Donald G. Marquis. "Research on the Human Factor in the Transfer of Technology." In William H. Gruber and Donald G. Marquis (eds.). *Factors in the Transfer of Technology.* Cambridge: MIT Press, 1969.

Hagstrom, Warren O. *The Scientific Community.* New York: Basic Books, 1965.

Harari, Roland. "The Long March of Chinese Science." *Science Journal* (April 1968), pp. 78-84.

Herrera, Amilcar. "Social Determinants of Science in Latin America: Explicit Science Policy and Implicit Science Policy." *The Journal of Development Studies* IX, 1 (October 1972), pp. 19-38.

Hoffmann, Charles. *Work Incentive Practices and Policies in the People's Republic of China, 1953-1965.* Albany: State University of New York Press, 1967.

Hough, Jerry. "The Prerequisites of Areal Deconcentration: The Soviet Experience." In James J. Heaphey (ed.). *Spatial Dimensions of Development Administration.* Durham, North Carolina: Duke University Press, 1971.

Illinois Institute of Technology Research Institute, *Technology in Retrospect and Critical Events in Science.* Volume I. (prepared for: The National Science

Foundation under contract NSF-C535). December 15, 1968.

"Industrial Development in China: A Return to Decentralization." *Current Scene* VI, 22 (December 20, 1968), pp. 1-17.

Ishikawa, Shigeru. "A Note on the Choice of Technology in China." *The Journal of Development Studies* IX, 1 (October 1972), pp. 161-186.

Jacob, Philip E. "Values and Public Vitality: The Political Dynamics of Community Activeness." In International Studies of Values in Politics. *Values and the Active Community*. New York: The Free Press, 1971, pp. 3-39.

Johnson, Chalmers (ed.). *Ideology and Politics in Contemporary China*. Seattle: University of Washington Press, 1973.

Jones, Graham. *The Role of Science and Technology in Developing Countries*. New York: Oxford University Press, 1971.

Kahn, A.Z.M. Obaidullah. "Class Struggle in Yellow Sandhill Commune." *The China Quarterly* 51 (July-September 1972), pp. 535-546.

Klochko, Mikhail A. *Soviet Scientist in Red China*. (trans. Andrew MacAndrew). New York: Praeger, 1964.

Kordon, Bernardo. "Kuo-Mo-jo Discusses the Cultural Revolution." *Clarin* (Santiago). February 9, 1969. In *JPRS* 47685.

Kornhauser, William. *Scientists in Industry: Conflict and Accommodation*. Berkeley: University of California Press, 1962.

Kuhn, Thomas S. *The Structure of Scientific Revolutions*. Chicago: The University of Chicago Press, 1962.

Kuo, Leslie T.C. *The Technical Transformation of Agriculture in Communist China*. New York: Praeger, 1972.

Kurakov, I.G. *Science, Technology and Communism, Some Questions of Development* (trans. Carin Dedijer). Oxford: Pergamon Press, 1966.

Kwok, Danny Wynn Ye. *Scientism in Chinese Thought, 1900-1950*. New Haven: Yale University Press, 1965.

Landau, Martin. "Linkage, Coding and Intermediacy: A Strategy for Institution-Building." *Journal of Comparative Administration*. Vol. 2, No. 4 (February 1971), p. 408.

Lee, Hahn-Been. "Developmentalist Time, Developmental Entrepreneurs, and Leadership in Developing Countries." In Dwight Waldo (ed.). *Temporal Dimensions of Development Administration*. Durham: Duke University Press, 1970.

Lee III, Rensselaer W. "Ideology and Technical Innovation in Chinese Industry, 1949-1971." *Asian Survey* XII. 8 (August 1972), pp. 647-661.

_______. "The Politics of Technology in Communist China." In Chalmers Johnson (ed.). *Ideology and Politics in Contemporary China*. Seattle: University of Washington Press, 1973.

Levy, Marion J. *Modernization and the Structure of Society*. Princeton: Princeton University Press, 1966.

Lewis, John Wilson. *Leadership in Communist China*. Ithaca, New York: Cornell University Press, 1963.

_______. (ed.). *The City in Communist China.* Stanford: Stanford University Press, 1971.

_______. (ed.). *Party Leadership and Revolutionary Power in China.* Cambridge: Cambridge University Press, 1970.

Lifton, Robert Jay. *Thought Reform and the Psychology of Totalism.* New York: W.W. Norton and Co., Inc., 1963.

Lindbeck, John M.H. (ed.). *China: Management of a Revolutionary Society.* Seattle: University of Washington Press, 1971.

_______. "Organization and Development Science." In Sidney H. Gould (ed.). *Sciences in Communist China.* Washington: American Association for the Advancement of Science, 1961, pp. 36-42.

Liu, Alan P. *Communications and National Integration in Communist China.* Berkeley: University of California Press, 1971.

Lubkin, Gloria B. "Physics in China." *Physics Today* XXV, 12 (December 1972), pp. 23-28.

Luzhnaya, N.P. "Scientific Institutes of the People's Republic of China." *Zhurnal Neorganicheskoy Khimii (Journal of Inorganic Chemistry)*, Vol. III, No. 2 (1958). In *JPRS* 2568.

McClelland, David C. "Motivational Patterns in Southeast Asia with Special Reference to the Chinese Case." *The Journal of Social Issues* XIX, 1 (January 1963), pp. 6-19.

Macioti, Manfredo. "Hands of the Chinese." *New Scientist and Science Journal* (June 10, 1972), pp. 636-638.

Maddox, John. "Choice and the Scientific Community." *Minerva* II, 2 (Winter 1964), pp. 141-159.

Mansfield, Edwin. *Technological Change.* New York: W.W. Norton and Co., Inc., 1971.

Marquis, Donald G. and Thomas J. Allen. "Communications Patterns in Applied Technology." *American Psychologist* Vol. 21, No. 11 (November 1966), p. 1055.

Meisner, Mitch. "The Shenyang Transformer Factory—A Profile." *The China Quarterly* 52 (October-December 1972), pp. 717-736.

Meltzer, Leo. "Scientific Productivity in Organizational Settings." *Journal of Social Issues* 12 (1956), pp. 32-40.

Mendelssohn, K. "Science in China." *Nature* 215 (July 1, 1967), pp. 10-12.

Merton, Robert K. *Science, Technology and Society in 17th Century England.* New York: Howard Fertig, 1970.

Moravcsik, Michael J. "Aspects of Science Development." Organisation for Economic Co-operation and Development, *Management of Research and Development.* Paris, 1972.

Morison, Elting E. *Men, Machines, and Modern Times.* Cambridge: MIT Press, 1966.

Morton, J.A. *Organizing for Innovation.* New York: McGraw-Hill, 1971.

Morton, Louis. "War, Science and Social Change." In Kalman H. Silvert (ed.).

The Social Reality of Scientific Myth. New York: American Universities Field Staff, Inc., 1969.

Nayudamma, Y. "Decentralized Management of R & D in a Developing Country." *Minerva* XI, 1 (October, 1973), pp. 516-536.

_______. "Promoting the Industrial Application for Research in an Underdeveloped Country." *Minerva.* V, 3 (Spring, 1967), pp. 323-339.

Needham, Joseph. *The Grand Titration.* Toronto: University of Toronto Press, 1969.

_______. "Science and Society in East and West." In Maurice Goldsmith and Alan Mackey (eds.). *Science and Society.* New York: Simon and Schuster, 1964.

Needham, Joseph and Dorothy Needham (eds.). *Science Outpost: Papers of the Sino-British Science Cooperative Office.* London: The Pilot Press, 1948.

Oldham, C.H.G. "Science and Education." *Bulletin of the Atomic Scientists* (June 1966), pp. 41-50.

_______. "Science Travels the Mao Road." *The Listener* (August 22, 1968), pp. 17-18.

"Organization et Developpement de la Science en Republique Populaire de Chine." *Notes et Etudes Documentaires*, 3255 (18 janvier, 1966). Paris La Documentation Francaise, Secretariat General du Government, 13.

Organisation for Economic Co-operation and Development. *Analytical Methods in Government Science Policy.* Paris, 1972.

_______. *The Conditions for Success in Technological Innovation.* Paris, 1971.

_______. *Management of Research and Development.* Paris, 1972.

_______. *Science Growth and Society: A New Perspective.* Paris, 1971.

Orleans, Leo A. "China's Science and Technology: Continuity and Innovation." Washington, D.C.: U.S. Congress. Joint Economic Committee. Joint Committee Print, 1972.

_______. "How the Chinese Scientist Survives." *Science.* 177 (September 8, 1972), pp. 864-866.

_______. *Professional Manpower and Education in Communist China.* Washington: National Science Foundation, 1960.

_______. "Research and Development in Communist China." *Science* 157 (July 28, 1967), pp. 392-400.

_______. "Research and Development in Communist China: Mood, Management, and Measurement." In U.S. Congress Joint Economic Committee. *An Economic Profile on Mainland China.* New York: Praeger Publishers, 1968.

Peake, Cyrus H. "Some Aspects of the Introduction of Modern Science into China." *Isis* XXII (1934-1935), p. 183.

Pelzel, John C. "Economic Management of a Production Brigade in Post-Leap China." In W.E. Willmott (ed.). *Economic Organization in Chinese Society.* Stanford: Stanford University Press, 1972.

Perkins, Dwight H. "Incentives and Profits in Chinese Industry." *Current Scene* IV, 10 (May 15, 1966), pp. 1-10.

_______. "Industrial Planning and Management." In Alexander Eckstein, Walter Galenson, and Ta-chung Liu (eds.). *Economic Trends in Communist China.* Chicago: Aldine, 1968.

_______. "Mao Tse-Tung's Goals and China's Economic Performance." *Current Scene* IX, 1 (January 7, 1971), pp. 1-15.

Price, Derek J. De Solla. "Is Technology Historically Independent of Science? A Study in Statistical Historiography." *Technology and Culture.* 6 (1965).

Price, William J. and Lawrence W. Bass. "Scientific Research and the Innovative Process." *Science* 164, 3881 (May 16, 1969), pp. 802-806.

Prybyla, Jan S. "The Chinese Economy: Priorities and Policy." *Current Scene* X, 6 (June 10, 1972), pp. 1-10.

_______. *The Political Economy of Communist China.* Scranton, Pennsylvania: International Textbook Co., 1970.

Pye, Lucian. *The Spirit of Chinese Politics.* Cambridge: MIT Press, 1968.

Rahmen, A., N. Sen, and N.G. Rajagopal. *Scientific Societies in India.* New Delhi: Survey and Planning of Scientific Research Unit, Council of Science and Industrial Research, 1965.

Rawski, Thomas G. "Recent Trends in the Chinese Economy." *The China Quarterly* 53 (January-March 1973), pp. 1-33.

"Reasons For the Failure of Mao Tse-Tung's Attempts to Achieve Socioeconomic Reforms in Rural China." *Literaturnaya Gazetaa (Literary Gazette)*, No. 27. July 5, 1967. In *JPRS* 42097.

Richman, Barry M. "Economic Development in China and India: Some Conditioning Factors." *Pacific Affairs* XLV, 1 (Spring 1972), pp. 75-91.

_______. *Industrial Society in Communist China.* New York: Vintage, 1969.

_______. "Managerial Opposition to Product Innovation in Soviet Union Industry." *California Management Review* VI, 2 (Winter 1963), pp. 11-26.

Riskin, Carl. "Local Industry and Choice of Techniques in Planning of Industrial Development in Mainland China." United Nations Industrial Development Organization, *Planning for Advanced Skills and Technologies.* Vienna, 1969.

_______. "Small Industry and the Chinese Model of Development." *The China Quarterly* 46 (April-June 1971), pp. 245-273.

Rogers, Everett M. and F. Floyd Shoemaker. *Communication of Innovations: A Cross Cultural Approach.* New York: The Free Press, 1971.

Sanders, Royden C. Jr. "Interface Problems Between Scientists and Others in Technically Oriented Companies." In Karl B. Hill (ed.). *The Management of Scientists.* Boston: Beacon Press, 1964.

Schurmann, Franz. *Ideology and Organization in Communist China.* 2d ed. Berkeley: University of California Press, 1968.

_______. "Politics and Economics in Russia and China." In Donald W. Treadgold (ed.). *Soviet and Chinese Communism: Similarities and Differences.* Seattle: University of Washington Press, 1967.

"Science and Research, Summer 1964." *China News Analysis.* No. 536 (October 9, 1964), pp. 1-7.

"Science in Underdeveloped Countries." Statement made by the Sussex Group to the Advisory Committee on the Application of Science and Technology to Development. *Minerva*. 9, 1 (January 1971).

"Scientific and Technical Education in China." *Ajiya Kenkyu (Asia Research)*. September 10, 1963. In *JPRS* 22237.

"Scientists, 1957-1967." *China News Analysis*. No. 696 (February 16, 1968), pp. 1-7.

Shih, Joseph Anderson. "Science and Technology in China." *Asian Survey* XII, 8 (August 1972), pp. 662-675.

Signer, Ethan and Arthur W. Galston. "Education and Science in China." *Science* 175, 4017 (January 7, 1972), pp. 15-23.

Sigurdson, Jon. "Rural Industry–A Traveller's View." *The China Quarterly* 50 (April-June 1972), pp. 315-332.

Skinner, G. William. "Marketing and Social Structure in Rural China, Part I." *The Journal of Asian Studies*. XXIV, 1 (November 1964), pp. 3-44.

_______. "Marketing and Social Structure in Rural China, Part II." *The Journal of Asian Studies*. XXIV, 2 (February 1965), pp. 195-228.

_______. "Marketing and Social Structure in Rural China, Part III." *The Journal of Asian Studies* XXIV, 3 (May 1965), pp. 363-400.

Skinner, G. William and Edwin A. Winckler. "Compliance Succession in Rural Communist China: A Cyclical Theory." In Amitai Etzioni. *A Sociological Reader on Complex Organizations*. New York: Holt Rinehart and Winston Inc., 1969.

Solomon, Richard H. "Communications Patterns and the Chinese Revolution." *The China Quarterly*. 32 (October-December 1967), pp. 88-110.

_______. "Mao's Effort to Reintegrate the Chinese Polity: Problems of Authority and Conflict in Chinese Social Processes." In A. Doak Barnett (ed.). *Chinese Communist Politics in Action*. Seattle: University of Washington Press, 1969.

_______. *Mao's Revolution and the Chinese Political Culture*. Berkeley: University of California Press, 1971.

_______. "On Activism and Activists: Maoist Conceptions of Motivation and Political Role Linking State to Society." *The China Quarterly* 39 (July-September 1969), pp. 76-114.

_______. "Parochialism and Paradox in Sino-American Relations." *Asian Survey* VII, 12 (December 1967), pp. 831-850.

Stewart, Rosemary. "Managers Under Mao." *Management Today* (April 1967), pp. 66-71.

Storer, Norman W. *The Social System of Science*. New York: Hall Rinehart and Winston, Inc., 1966.

Suttmeier, Richard P. *The Chinese Academy of Sciences: Institution Building in a Research Oriented Organization*. Unpublished doctoral dissertation. Department of Government, Indiana University, 1969.

_______. "Chinese Scientific Societies and Chinese Scientific Development." *The Developing Economies* XL, 2 (Autumn 1973).

_______. "Party Views of Science: The Record From the First Decade." *The China Quarterly* 44 (October-December 1970), pp. 146-168.

Tang, Anthony M. "Policy and Performance in Agriculture." In Alexander Eckstein, Walter Galenson, and Ta-chung Liu (eds.). *Economic Trends in Communist China.* Chicago: Aldine, 1968.

Teich, Albert H. (ed.). *Technology and Man's Future.* New York: St. Martin's Press, 1972.

Teng Ssu-yu and John K. Fairbank (eds.). *China's Response to the West: A Documentary Survey, 1839-1923.* New York: Atheneum, 1966.

Thompson, Victor A. *Bureaucracy and Innovation.* University, Alabama. The University of Alabama Press, 1968.

Toulmin, Stephen. "The Complexity of Scientific Choice II: Culture, Overheads or Tertiary Industry?" *Minerva* IV, 2 (Winter 1966), pp. 155-169.

_______. "The Complexity of Scientific Choice: A Stocktaking." *Minerva* II, 3 (Spring 1964), pp. 343-359.

Townsend, James R. *Political Participation in Communist China.* Berkeley: University of California Press, 1968.

Uchida, Genko. "Technology in China." *Scientific American* 215, 5 (November 1966), pp. 37-45.

Union Research Institute. *Who's Who in Communist China.* Hong Kong: Union Research Institute, 1966.

U.S. Department of Commerce. *Technological Innovation: Its Environment and Management.* Washington, 1967.

Varsovsky, Oscar. "Scientific Colonialism in the Hard Sciences." *American Behavioral Scientist* X (June 1967), pp. 22-23.

Vogel, Ezra F. "Politicized Bureaucracy: Communist China." In Fred W. Riggs (ed.). *Frontiers of Development Administration.* Durham, North Carolina: Duke University Press, 1970.

_______. "Voluntarism and Social Control." In Donald W. Treadgold (ed.). *Soviet and Chinese Communism.* Seattle: University of Washington Press, 1967.

Vollmer, Howard M., Todd R. La Porte, William C. Pedersen, and Phyllis A. Langton. *Adaptations of Scientists in Five Organizations: A Comparative Analysis.* Menlo Park, California: Stanford Research Institute, 1964.

Vucinich, Alexander. *The Soviet Academy of Sciences.* Stanford, California: Stanford University Press, The Hoover Institute Studies. Series E: Institutions. No. 3, 1956, p. 22.

Watkins, Ralph J. (ed.). *A Directory of Selected Scientific Institutions In Mainland China.* Stanford: The Hoover Institution Press, 1971.

Weinberg, Alvin M. "Criteria for Scientific Choice." *Minerva* I, 2 (Winter 1963), pp. 159-171.

_______. "Criteria for Scientific Choice II: The Two Cultures." *Minerva* III, 1 (Autumn 1964), pp. 3-14.

Whyte, Martin King. "Bureaucracy and Modernization in China: The Maoist Critique." *American Sociological Review* XXXVIII, 2 (April 1973), pp. 149-163.

Wu Yuan-li and Robert B. Sheeks. *The Organization and Support of Scientific Research and Development in Mainland China.* New York: Praeger, 1970.

Yang, Chen-ning. "Education and Scientific Research in China." *Asia* XXVI (Summer 1972), pp. 74-84.

Yeh, K.C. "Soviet and Communist Chinese Industrialization Strategies." In Donald W. Treadgold (ed.). *Soviet and Chinese Communism.* Seattle: University of Washington Press, 1967.

Yu, Frederick. "With Banners and Drums: The Mass Campaign in China's Drive for Development." *Current Scene* IV, 9 (May 1, 1966), pp. 1-13.

_______. *Mass Persuasion in Communist China.* New York: Praeger, 1964.

Ziman, J.M. *Public Knowledge.* Cambridge: Cambridge University Press, 1968.

Index

Index

About the Author

Richard P. Suttmeier is Assistant Professor of Government at Hamilton College, Clinton, New York. He received the B.A. from Dartmouth College and the Ph.D. in political science from Indiana University. From 1969 to 1970 he served as an American Society for Public Administration Fellow in the National Aeronautical and Space Administration; from 1972 to 1973, he was a Visiting Fellow at the Technology and Development Institute of the East-West Center, Honolulu, Hawaii.